Why do buses come in threes?

Why do buses come in threes?

The hidden mathematics of everyday life

Rob Eastaway
and
Jeremy Wyndham

Illustrations by Barbara Shore

JOHN WILEY & SONS, INC.
New York • Chichester • Weinheim • Brisbane • Singapore • Toronto

This book is printed on acid-free paper. ♾

Published by John Wiley & Sons, Inc.
Published simultaneously in Canada.

First published in Great Britain in 1998 by Robson Books Ltd, Bolsover House, 5-6 Clipstone Street, London, W1P 8LE, England.

Illustrations by Barbara Shore

Library of Congress Cataloging-in-Publication Data

Eastaway, Robert.
Why do buses come in threes? : the hidden mathematics of everyday life / Rob Eastaway and Jeremy Wyndham ; illustrations by Barbara Shore.
p. cm.
Includes bibliographical references and index.
ISBN 0-471-34756-6 (cloth: alk. paper)
ISBN 0-471-37907-7 (paper: alk. paper)
1. Mathematics Popular works. I. Wyndham, Jeremy. II. Title.
QA93.E18 1999
510—dc21 99-21915

Printed in the United States of America

10 9 8 7

Contents

Foreword

We did not invent mathematics, we discovered it. It exists in every aspect of our lives, serious or light-hearted, momentous or trivial. The subject is often misunderstood and unreasonably feared, yet it is simpler and more logical than any language. When we gaze up into the skies at night, wondering at the beauty and inaccessibility of the stars, when we get into a bath displacing (in my case quite a lot of) water, when we read the football results or toss a coin, knowledge of mathematics and its related disciplines can help us enjoy and understand, even predict and prepare for the future.

My three greatest enthusiasms as a child were for cricket, pop music and astronomy. All three, though I did not realise it at the time, came about because of statistics – the batting averages, the pop charts, the sizes and distances of the planets. That common strand of numbers to these apparently unconnected topics launched me into three lifelong passions, and there have been plenty of other occasions when figures have been the basis for a new interest, although my lack of success at roulette and with bookmakers over the years has occasionally made me wish this had not always been so.

The most beautiful pieces of music can be broken down mathematically – all notes have a numerical relationship to each other, vibrating in harmony, unison or discord – the purer and more straightforward the mathematical connection, the sweeter the sound. I am not suggesting that Mozart or Bob Dylan should only be listened to with a calculator to hand, and I doubt whether either created his works of emotional genius with oscillations per minute in his mind, but if some higher being did not, I should be very surprised.

Rob Eastaway and Jeremy Wyndham claim that this book is fun and they are dead right. From potato crisps to snooker balls, from card tricks to insurance, from code-breaking to bus-waiting, everything herein reminds us how mathematics rules and enhances our existence.

Tim Rice

Acknowledgements

Much of the inspiration for this book came from the work of Martin Gardner, who has done so much to popularise mathematics in the last forty years. We would also like to give special thanks to David Wells for contributing so many ideas; to David Singmaster for access to his library and his own encyclopaedic knowledge of mathematical recreations; and to Malcolm Field for his mathematical expertise.

We are indebted to those who painstakingly looked through the early drafts of this manuscript, especially to Martin Daniels, Steve Barsky, David Flavell and Sarah Wyndham. Thanks also to Jack Eastaway, Barbara Brown, Tony Taylor, Harold Lind, Jo Lehrman and Sam Banks.

In addition, Lionel Titman, Tim Jones, Craig Dibble, Hugh Jones, Darren Nicholls, Dennis Sherwood, Paul Harris, Cheryl Kramer, Richard Hamill, Chris Healey, Susan Blackmore, Martin Turner, Helen Nicol, Emma Rushton and Michael Balle provided us with helpful contributions.

Charlotte Howard deserves a special mention for her encouragement, as does everyone at Robson Books and John Wiley for their enthusiastic support.

And finally, thank you Elaine and Sarah for being such enthusiastic and understanding supporters throughout.

Introduction

Mathematics is fascinating, beautiful, sometimes even magical. It is relevant to just about everything that we do, and full of topics suitable for the most stimulating dinner party conversations. That may not be the popular view, but it is certainly ours and we hope it might be yours too. Math has had a bad press for far too long, and it is time to put the case for the defence. This book is for anybody who is interested in reminding themselves – or discovering for the first time – that mathematics is an essential part of our lives.

Have you ever asked yourself why it is that buses come in threes? As a child, did you share the frustration of not finding a four-leafed clover? When you bump into an old friend miles from home, do you smile to yourself in amazement that coincidences like this can happen? Occurrences like these interest everyone, and the explanations behind all of them are mathematical. But maths doesn't just answer questions. It also provides new insights and it stimulates curiosity. Gambling, travelling, dating, eating, even deciding whether or not to run when it's raining, all involve elements of maths.

Books about popular and recreational mathematics can often seem abstract and inaccessible to those who have lost touch with the subject since their schooldays. We have tried to bring maths back into real, everyday life. That's why every chapter begins with a question that might occur to anybody. The choice of material reflects our personal interests rather than any grand logical scheme. Some bits are easy reading, others require a little more thought, but whatever your mathematical ability there will be plenty here for you.

Dotted through the book you will find practical uses for probability theory, as well as surprising applications of tangents, Fibonacci series, pi, matrices, Venn diagrams, prime numbers and more. We hope you find these subjects as thought-provoking and stimulating as we do. Above all, we want you to enjoy it.

1

WHY CAN'T I FIND A FOUR-LEAFED CLOVER?

Links between nature and mathematics.

One of the magical adventures of childhood is searching for a four-leafed clover. It's the next best thing to hunting for the pot of gold at the end of the rainbow. Unfortunately, both of these quests usually end in disappointment. It is easy to give up on the rainbow's gold because the rainbow has usually disappeared before the child's curiosity, but the hunt for the clover is much more frustrating. It seems perfectly reasonable that somewhere there should be one with four leaves. So why does nature so rarely deliver?

Next time you are out in the garden or in the countryside, take a little time to study the flowers. You'll find that the commonest number of petals on a flower is five. Buttercups, mallow, pansies, primroses, rhododendrons, tomato blossom, geraniums . . . this is just a sample from the large number of flower families that have picked the number five. Even a flower that appears to have ten petals, such as red campion, has five petals each subdivided into two.

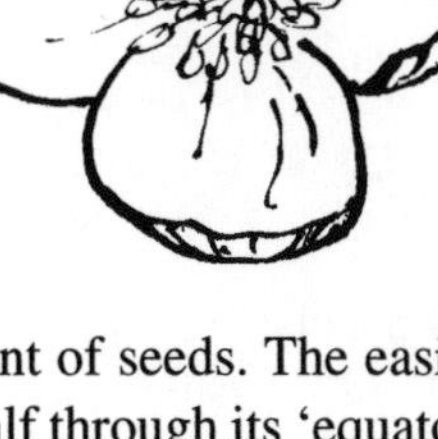

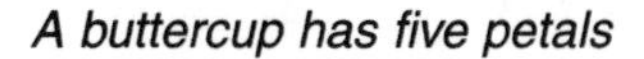

A buttercup has five petals

Fives also appear in the arrangement of seeds. The easiest way to find a pattern of five is to cut open an apple. If it is cut in half through its 'equator' (normally apples are cut from 'pole to pole' down the core) you will find the seeds arranged in a beautiful five-pointed star. It works with pears, too.

Why is there this odd number in plants, when in animals it is *even* numbers that are so common? (Legs usually come in twos, fours or sixes, for example.) Why choose five petals over the more symmetrical four or six?

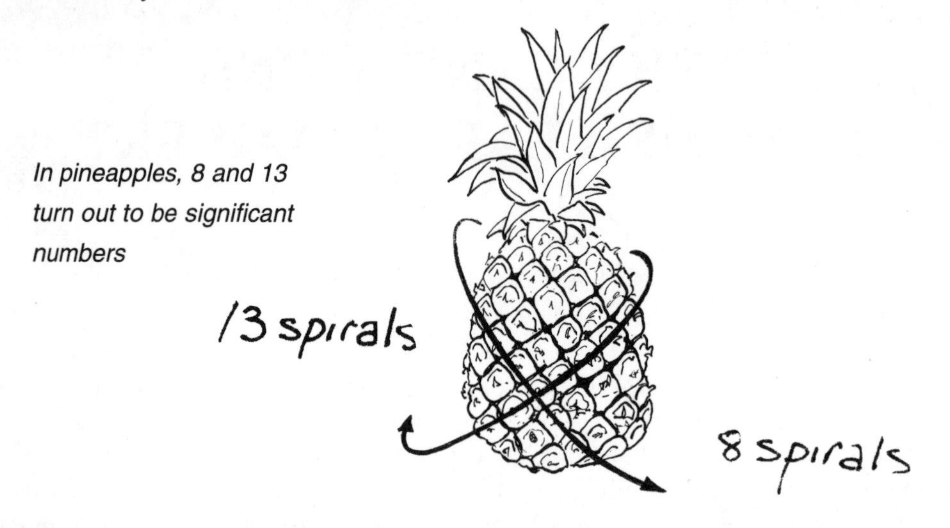

In pineapples, 8 and 13 turn out to be significant numbers

Further investigation leads to other numbers in plants that appear with curious frequency. Examine a pineapple or a pine cone and you will see that it has spiral rows of scales running from top to bottom. Two of these spirals are particularly easy to pick out: one runs clockwise, the other anticlockwise. The number of rows of spirals is usually 8 and 13 in a pineapple, while in a cone there might typically be 13 and 21 or 21 and 34. In a sunflower you will also find clockwise and anti-clockwise spirals, this time in the florets running from the centre of the flower outwards. The number of clockwise and anti-clockwise spirals will often be 34 and 55 or 55 and 89.

Those who have carried out the painstaking job of counting petals on a wide variety of flowers claim that 8,13, 21, 34 and 55 are more common than their neighbouring numbers. A flower has eight petals more often than seven or nine.

It is not a coincidence that some numbers appear more often than others. Indeed there is an intriguing connection between petals, leaves and pine cones, and an area of mathematics that has been a source of fascination for hundreds of years.

Fibonacci's Series

An Italian called Leonardo Fibonacci (1170-1240) gave his name to a simple number series. This series starts with 1 and 1, and each subsequent number in the series is calculated by adding the previous two. Fibonacci's series goes like this:

1, 1, 2, 3, 5, 8, 13, 21, 34, 55 . . . and so on.

Fibonacci originally produced the series when he was working out how many rabbits he would have if they bred at a particular rate. Yet the Fibonacci series has turned out to be one that has a link with nature that is far more profound than simple rabbit numbers. You may already have noticed that all of the petal and scale numbers mentioned earlier were Fibonacci numbers. Leaves, too, most often come in twos, threes and fives. Hence, clovers fit the pattern by usually coming with three leaves and not four.

But why is it that Fibonacci numbers crop up so often in plants?

It all comes down to the link between the Fibonacci series and a special number that ancient civilisations believed to have divine and mystical properties. That special number is the golden ratio.

The golden ratio

The golden ratio, or ϕ (phi) is $(\sqrt{5} + 1)/2$. This works out at about 1.618. As a number it may not have sent your pulse racing, yet its importance in nature is profound. This ratio belongs to a particular rectangle with a unique property.

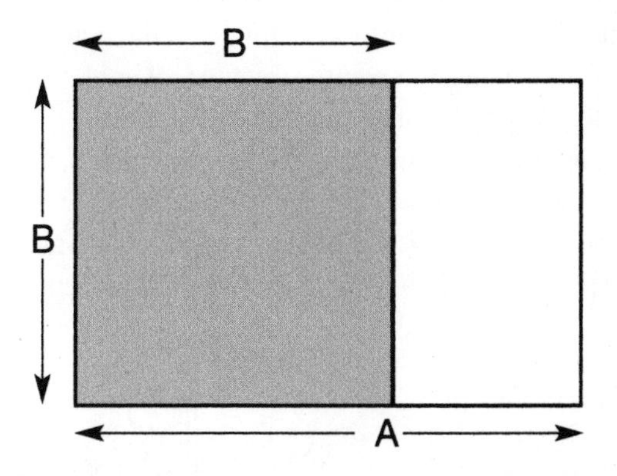

Here is a special rectangle of dimensions A x B. If you cut off a square BxB (as shown) then the remaining rectangle has exactly the same proportions as the original. This property is unique to a golden rectangle. The ratio of A to B is 1.618...., otherwise known as phi or ϕ.

ϕ doesn't just appear in a rectangle. It features in every pentagon and five-pointed star, which means you find it in apples too.

Take the star you found in the apple earlier on. You should find that the distance between the tips of the first and third points of the star is ϕ times the distance between adjacent tips. (At least, it would be with a perfect star and a precision ruler.)

And that isn't the end of ϕ's curious properties.

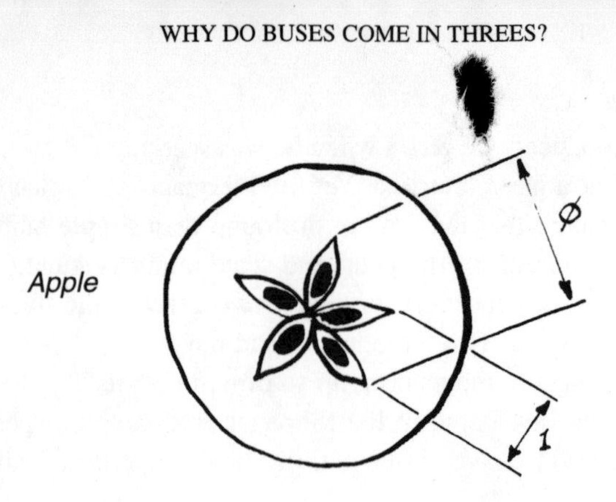

The ratio of any consecutive pair of numbers in the Fibonacci series is roughly ϕ, 3/2 = 1.5, 5/3 = 1.6 and so on. The further along the series, the closer the ratio of terms is to ϕ. By the time you get to 34/21, or 1.619, the ratio is already within 0.1 per cent of the precise value. Fibonacci and the golden ratio are intimately linked to each other.

Now let's return to plants. In many plants, you will notice that there are individual leaves which sprout from the stem. These leaves usually stick out from the stem at different angles, and as you move up the stem the leaves form a spiral. The angle by which each leaf is rotated from the previous one is usually between 137 and 139 degrees. A quick experiment in the garden confirmed this, by the way. The first weed plucked from a flowerbed had nine leaves separated by just over three revolutions. The average angle between each leaf worked out at roughly 139 degrees.

What is the significance of this angle? As will emerge, it is connected to ϕ, but why? It all comes down to what happens to a plant in its infancy. Every leaf and petal first appears as a tiny bud. The buds appear one at a time down the stalk. Each bud tries to position itself as far away from the previous ones as possible, almost like a repelling magnet. The reason *why* this

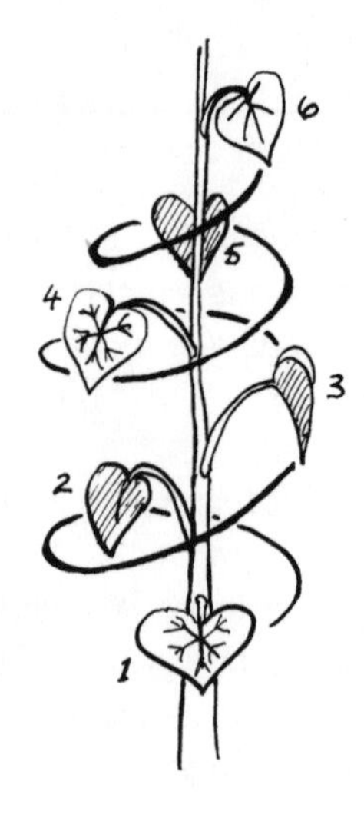

Leaves spiral up the stem of a garden weed

happens is probably that each bud wants as much space and light as it can get so that it can grow. To achieve this, the bud points itself at a different angle from its predecessors.

It just so happens that an angle related to ϕ is particularly well suited to keeping the buds as far from all their predecessors as possible. 360 degrees divided by ϕ is about 222.5 degrees. 222.5 degrees clockwise represents the same amount of turn as 137.5 degrees anticlockwise, and this is the angle that appears time and again in plants.

It also turns out that if each bud emerges 137.5 degrees rotated from the previous one, something interesting happens with the sixth bud, as the diagram shows.

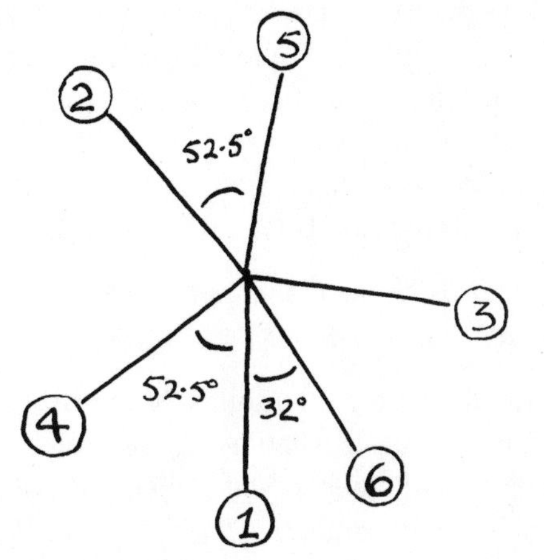

Angles between buds of a flower

Positions of the first six buds. Note that the fourth and fifth buds to appear are at least 52.5° clear of the buds above, but that the sixth is only 32.5° clear of the first.

The fourth and fifth buds are both at least 50 degrees out of alignment with each of their predecessors. However, the sixth bud to appear is only 32.5 degrees out of alignment with the first bud. If you like, the sixth bud is slightly in the shade of the first bud. At least, it is more shaded than any of the other five buds. This means that the sixth bud has slightly less access to sunlight and nutrients than the other buds, and this might just tip the balance on whether it grows or not. Could *this* be the reason why so many plants stop at five? Do many plants have a cut-off point programmed into them that inhibits the sixth bud from forming? It is a theory which has a certain charm to it, although nobody seems to understand the full story.

All of this has been merely an introduction to the intricate connections between Fibonacci, the golden ratio and the number five. What it shows, though, is that a plant's design may have as much to do with numbers as with its genes.

The link with plants is one reason why the golden ratio has been a source of fascination and reverence for so many centuries. Even the ancient Egyptians knew of the ratio, and the face of the Pyramid of Giza is made up from two halves of what is very nearly a golden rectangle.

There is, however, another shape which is even more closely linked with nature. It, too, contains a ratio with some mysterious qualities.

Pi and the circle

Circles are everywhere, in the fields, forests, oceans and the sky. Seeds, flower heads, eyes, tree trunks, rainbows and water droplets all contain circles. The planets also appear circular, and for a long time it was believed that they even moved in circles. (In fact planets move in ellipses, the circle being a special member of the ellipse family.)

The circle is commonplace because it is such an efficient shape, and because it is so easy to create. If a goat is tethered to a post in the middle of a field and the goat attempts to eat as much grass as possible, the shape of the grass it grazes will be a circle. If you have a fixed amount of fencing and want to enclose as large an area as possible, you will do quite well by creating a square, but you will enclose over 25 per cent more land if the fence makes a circle. Nature has a habit of finding optimal solutions – after all, it's had plenty of time to practise – and so it has exploited the circle to the full.

The ratio of the diameter of a circle to its circumference is known as pi (or π). Even in biblical times π was known to be about 3. According to the first Book of Kings, 7:23:

> *'And he made a molten sea, ten cubits from one brim to the other, and it was round all about ... and a line of thirty cubits did compass it round about'.*

In later times there were some who used this quote and the infallibility of the Bible to argue that π must be exactly 3. But, alas, neither dogma nor legislation can overcome the fact that π is a little less than $3^1/_7$. In fact it is an *irrational* number, which means its value can never be expressed as a single fraction that uses whole numbers.

Some odd facts about π...

- *A line down the middle of the number 113355 gives a ratio that is almost exactly 1/π. 113/355=1/3.1415929*
- *A useful mnemonic for π is: 'Can I find a trick recalling pi easily?'. The number of letters in each word gives the digit of π to seven decimal places: 3.1415926. And for 1/π: 'Can I remember the reciprocal' gives 0.318310, which is correct to six decimal places.*
- *There are many elegant series that can be used to create π. One of the simplest is: (1- 1/3 + 1/5 – 1/7 + 1/9 – 1/11...) x 4 although you need to go a long way into the series before you begin to get close to the right value.*
- *The ratio was first called π by William Jones in 1706. Jones was the son of a Welsh farmer from Anglesey.*
- *π also crops up in lots of important formulae that have no connections to circles at all – see later.*

Any natural phenomenon that involves circles will inevitably involve π as well. However, π can also crop up when the link with circles is less obvious. π occurs in time-keeping, for example. The time it takes for a gently swinging pendulum to go through one cycle is neatly summed up by this limerick:

If a pendulum's swinging quite free
Then it's always a marvel to me
That each tick plus each tock
Of the grandfather clock
Is 2π root L over g

$$2\pi\sqrt{\frac{L}{g}}$$

L is the length of the pendulum in metres and g is the rate of acceleration under gravity, which is about 9.8 metres per second per second on earth. The formula works on any planet, and since π is constant throughout the universe, a pendulum is a simple way of working out how strong the gravity is on a planet. A pendulum that is one metre long has a tick or tock every second when on earth, but on the moon each tick lasts two and a half seconds.

The 18th century biologist Georges Buffon made another intriguing discovery about π in the physical world. If a needle is dropped from a great height onto a flat surface on which parallel lines are drawn, the gap between each line being exactly one needle length, the chance that the needle will end up touching a line is exactly 2/π (that's about 64 per cent). A hundred years later, the mathematician Augustus de Morgan had one of his students test out this result. The student dropped a needle 600 times and it ended up touching a line 382 times, giving a value of π as 3.14 – suspiciously accurate, it might be said. But if you are trapped on a desert island and need to be able to calculate π as accurately as possible, you now have an extremely eccentric method for estimating it. Once you have found a stick and drawn the lines in the sand, all you will need to do is count. Mind you, in order to guarantee accuracy to three decimal places you need to drop the stick tens of thousands of times.

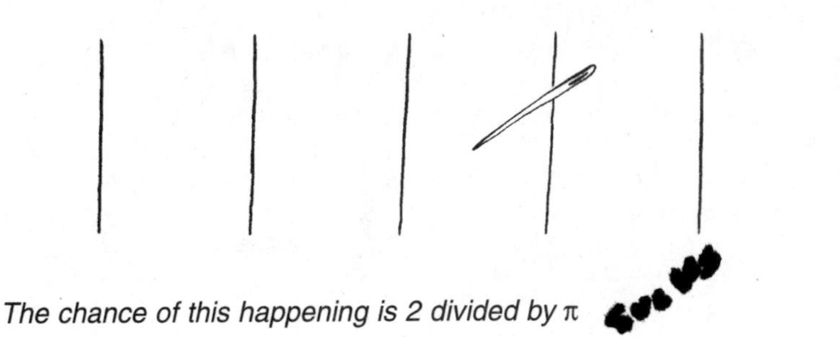

The chance of this happening is 2 divided by π

Wheels, and why animals don't have them

Although circles are important in nature, there is one place where they are notably absent. One of the most practical applications of circles is in one of man's greatest ever inventions, the wheel. Why are wheels circular? One reason is that circles have a uniform diameter, so the load being moved has a perfectly smooth ride. However, circles aren't the only shapes with constant diameter. Start with an equilateral triangle and from each corner draw an arc of a circle which meets the other two corners. The result is a shape with constant diameter. This shape could serve just as well as a roller, but it would be useless as a wheel because wheels need axles.

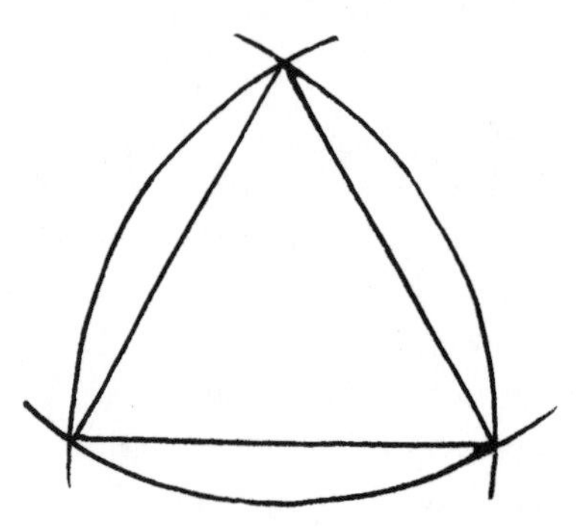

Curved triangle with constant diameter

The advantage that a circle has over other constant diameter shapes is that a circle has a centre which is equidistant from every point on its circumference. This means that it can

Why do 50 pence pieces have seven sides?

Any regular shape with an odd number of sides – a triangle, a pentagon, a heptagon and so on – can be rounded off to create a shape which has a constant diameter.

50 pence pieces have seven rounded sides so that the coin has constant diameter. This means it can be inserted into a slot machine in any orientation and it will still pass the 50 pence check. It wouldn't work if the coin had an even number of sides, since the diameter cannot be made constant. This is why the coins in any modern currency are either circular or have an odd number of sides.

Train wheel conundrum

Which part of a moving train is always stationary, and which part is always travelling in the reverse direction to the train itself?

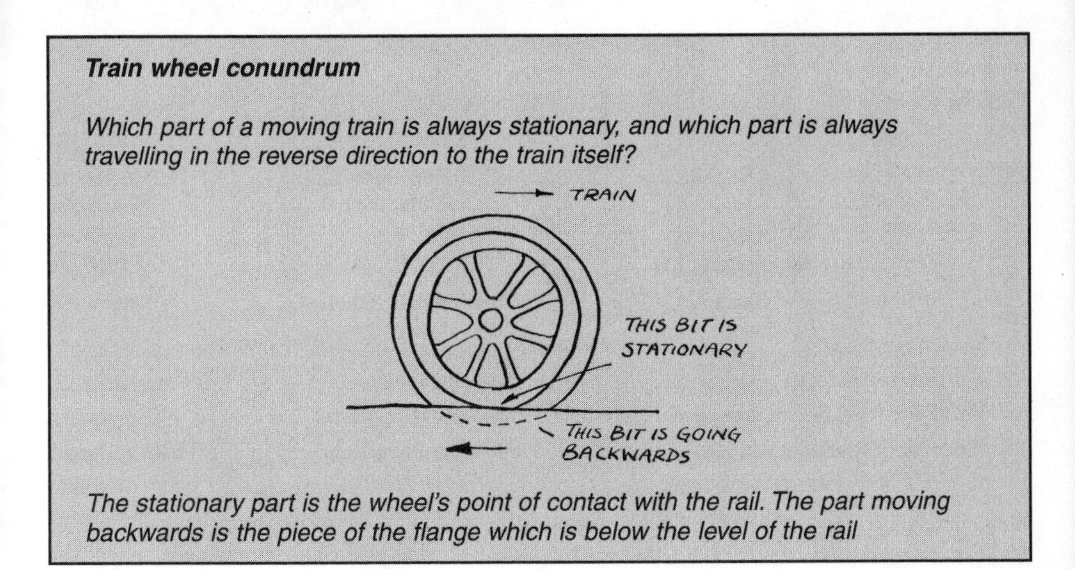

The stationary part is the wheel's point of contact with the rail. The part moving backwards is the piece of the flange which is below the level of the rail

have an axle which stays in the same position. On the triangular wheel, the axle would move up and down, making it impractical.

The biggest benefit of the wheel is that it saves energy. If a stone is pushed along the ground, the rubbing of the stone against the ground causes friction. Wheels, however, hardly rub against the ground at all. This is because the part of a moving wheel that is in contact with the ground is stationary for an instant.

But if wheels are so efficient, why do we not find them in animals? Animals seem to have discovered every other possible benefit from circles, so why don't we see kangaroos cruising around the Australian desert on two wheels instead of wasting all that energy hopping around on legs? The most likely reason is that wheels need axles; if wheels were part of the animal's body, the animal would need axles too. These axles would have to carry the sinews and blood vessels, and after a couple of rotations they would become horribly twisted.

Meanwhile the ability to roll easily, which is so useful in wheels, becomes a liability in eggs. Eggs usually have a circular cross-section because laying a square egg would bring tears to the eyes of any bird. However, it would be a disadvantage if eggs were *spherical* because they would then roll away from the mother too easily. If you gently roll a hen's egg on the floor, you will discover it comes back to you like a boomerang ... travelling in what looks suspiciously like a circle!

Honeycombs and hexagons

There is another situation in which circles are not ideal. A circle may have the most efficient ratio of circumference to enclosed area, but if circles are stacked on top of each other there is a lot of wasted space.

Stack of circles showing void space

Efficient packing and strength are particularly important qualities in nature, and no more so than in a beehive. If a stack of cylinders is arranged as in the diagram and compressed, the circles will form themselves into a tightly packed network of hexagons. It is no coincidence that this is the formation adopted by bees. No doubt bees would like to build circular cells for themselves, since circles are so strong, but bees also don't want to waste space or beeswax. Hexagons are the ideal compromise. The more sides that a regular polygon has (i.e. the higher the *order* of the polygon), the greater the area it can cover for a given perimeter length. Hexagons are better than squares and triangles, but not as good as heptagons, decagons or circles. However, the hexagon is the highest order of regular polygon that can be used to tile a floor without leaving any gaps. This helps to make hexagonal arrays the strongest structure for the least amount of material used.

And to complete nature's circle ...

Here is one final curiosity about hexagons in a beehive.

The cells in the honeycomb below have been labelled A, B, C, D ... Suppose a queen bee

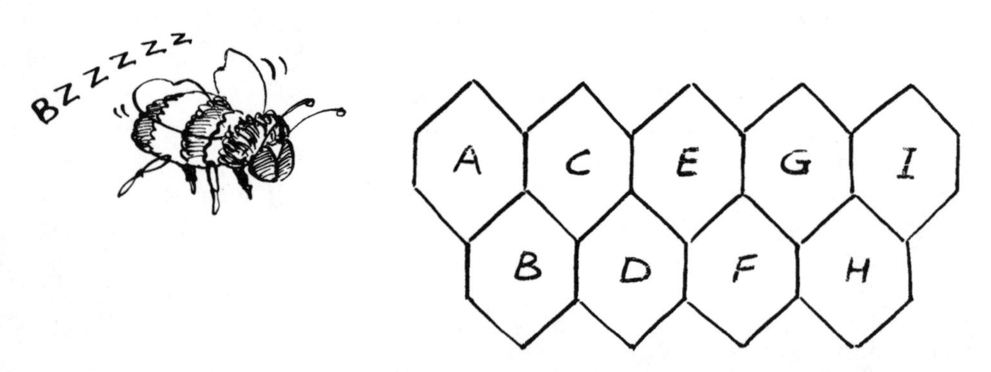

is visiting some of the cells in two rows of honeycomb and is always moving from left to right. The bee starts at cell A. There is only one possible path to cell B. There are two ways to cell C,moving left to right. One is to go A to C. The other is A-B-C. Now consider cell

D. The bee can either go A-B-D, A-C-D or A-B-C-D, in other words it has three possible paths. There are five possible paths to cell E, and eight to cell F.

There is a pattern emerging here: 1, 2, 3, 5, 8 ... we are back to Fibonacci again! It is only natural that everything should eventually come back to where it started.

Königsberg Bridges in 1763

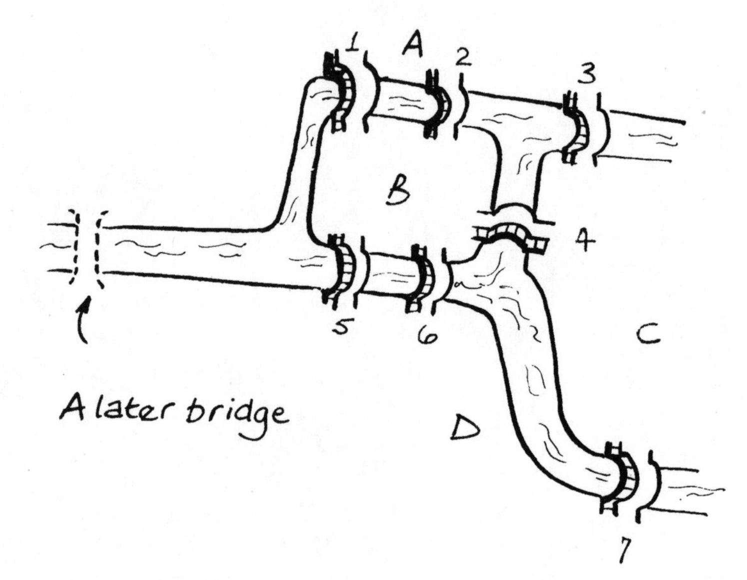

2

WHICH WAY SHOULD I GO?

From postmen to taxi drivers.

On the shores of the Baltic Sea wedged between Lithuania and Poland is a region of Russia known as the Kaliningrad Oblast. The city of Kaliningrad is, by all accounts, a bleak industrial port with shoddy grey apartment buildings built hastily after World War II, when the city had been obliterated first by Allied bombers and later by the invading Russian forces. Little remains of the beautiful Prussian city of Königsberg, as it was formerly known. This is sad not only for lovers of architecture but also for nostalgic mathematicians, because it was thanks to the layout of 18th century Königsberg that Leonhard Euler, one of the greatest mathematicians of all time, answered a puzzle which eventually contributed to two new areas of maths known as topology and graph theory.

Königsberg was built on the bank of the river Pregel. Seven bridges connected two islands and the banks of the river, as shown in the map opposite.

A popular pastime of the residents was to try to cross all the bridges in one complete circuit without crossing any of the bridges more than once. They knew how to make their own entertainment in those days.

This seemingly simple task proved to be more than tricky. Imagine the increasing frustration of a Königsberger taking a Sunday afternoon stroll. '1, 2, 3, 4, 5, 6 ... *Nein*! 1, 5, 7, 4, 2, 3 ... *Donner und Blitzen*!'. In fact nobody had been able to find a solution to the puzzle when Euler first heard of it and, intrigued by this, he set about proving that no solution was possible.

Euler analysed the problem by converting the map of the bridges into a *network* diagram.

Network of Königsberg bridges

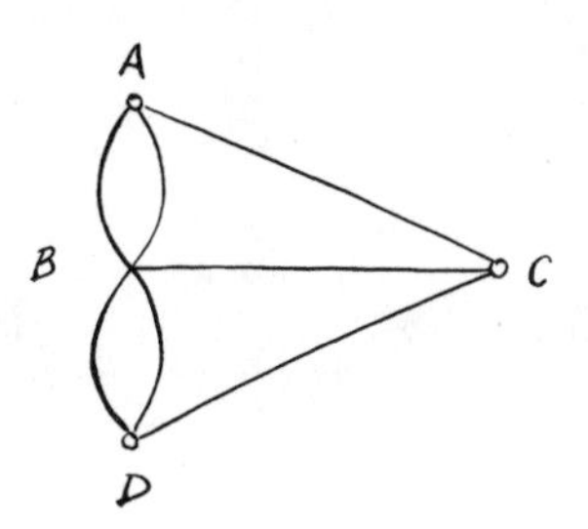

A network is a collection of dots connected by lines. The diagram on page 13 may not at first glance look like the map on page 12, but in mathematicians' terms they are exactly equivalent. That is to say, they are *topologically* equivalent (see the box below).

The dots labelled A B C D represent the north and south banks of the river (A and D) and the islands (B and C). The lines represent the paths or bridges by which A, B, C and D are connected. Two bridges connect A and B, two connect B and D, one connects B and C, one connects A and C and one connects C and D.

Euler described a dot or *node* as being either 'odd' or 'even'. A node is *odd* if there is an odd number of lines emerging from it, and *even* if there is an even number of lines. Euler studied many networks in addition to Königsberg, and proved:

A circuit in which each path is traversed only once can only be made when there are either no odd nodes or two odd nodes. In all other cases the network cannot be traced without doubling back.

Topology (or 'what does the London Underground really look like?')

Everybody has experienced 'topology' without necessarily realising it. There is no better example than the London Underground map, which is one of the great designs of modern times. No tourist has difficulty following it. 'Take the brown line to Oxford Circus, then change to the blue line and go two stops to Victoria'.

This neat network of straight lines and evenly spaced junctions bears almost no visual resemblance to the real layout of the city's underground lines. If you trace the underground on a normal map, it resembles an ungainly spider with straggly legs and almost nothing in the bottom right hand corner. But what matters to the tourist is the order of the stations and the intersections of the tube lines. It is as if the real map has been drawn on rubber and then squeezed and pulled until it fitted into a more convenient shape. That is topology.

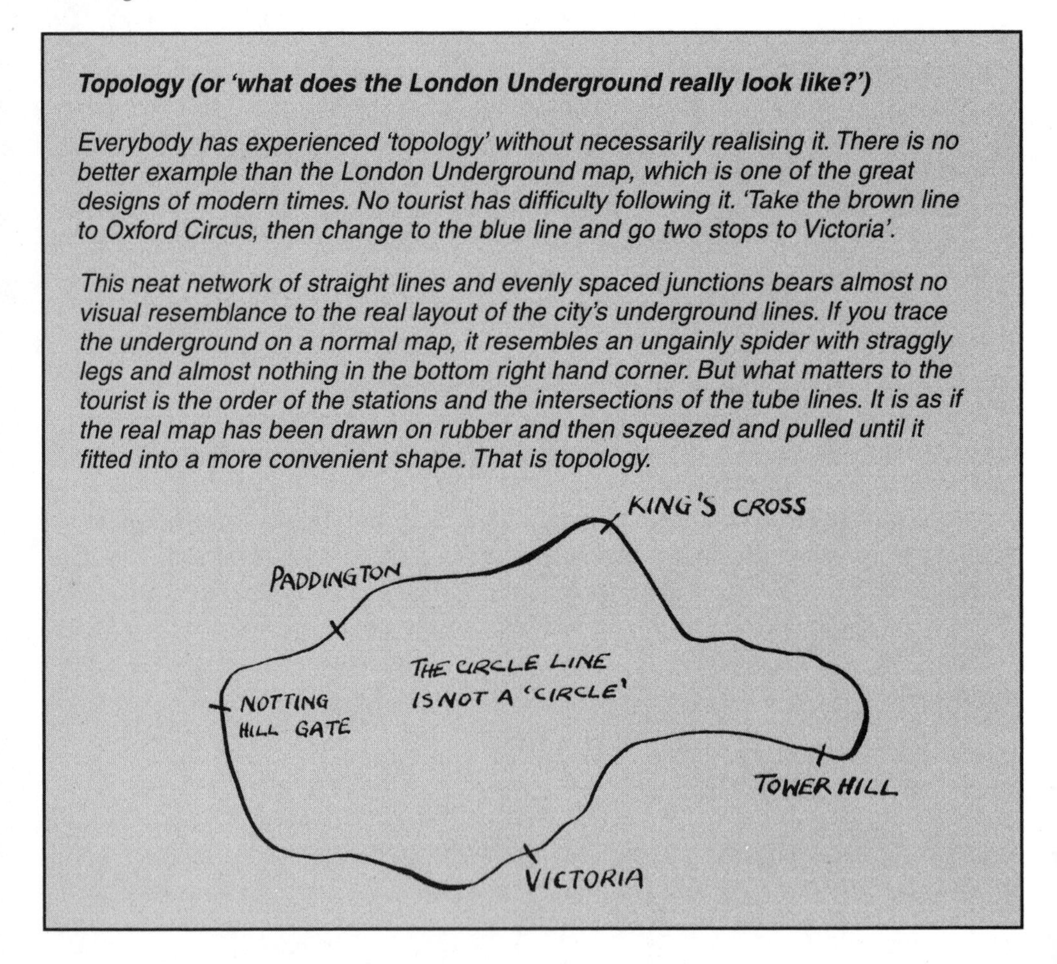

He also discovered: *If there are two odd nodes, the path through the circuit must start at one of the odd nodes and finish at the other one.*

At last there was a proof for the Königsberg puzzle. All four nodes A, B, C and D were odd, so by Euler's first rule there was no pathway that could possibly solve the original problem.

In the late 19th century an eighth bridge was built at the place marked on page 12 in the first diagram. It is not clear whether the city fathers built this bridge to change the city's puzzle for tourists or because there was a pressing traffic build-up, but the result was that Königsberg had been *Eulerised*. It was now possible to complete the tour without doubling back. The reason was that the number of odd nodes had been reduced to two, although by Euler's second rule this meant that anybody travelling the circuit would have to start at B and finish at C, or vice versa .

Sadly, in 1944 air raids obliterated most of the old bridges. However, from the maps made available since, it appears that five bridge crossings were rebuilt, leaving the centre of the city like this:

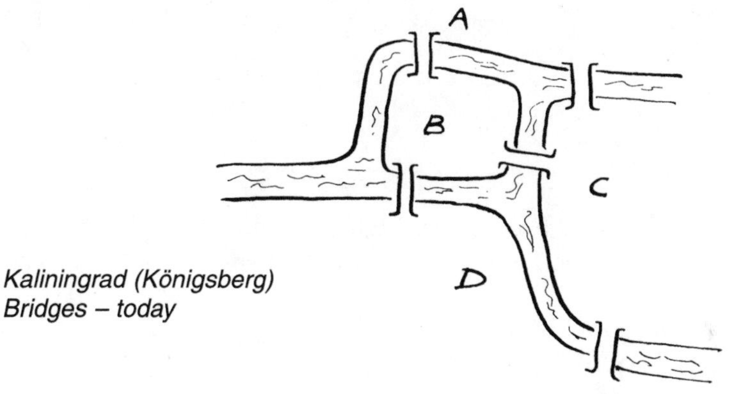

Kaliningrad (Königsberg) Bridges – today

It would seem, that once again Kaliningrad has been Eulerised, by taking the route B-C-A-B-D-C, for example. Did the Russians do this deliberately?

The postman only passes once

Königsbergers were only doing their Euler circuits for fun. There are many situations, however, where a tour without doubling back is of more serious concern.

For a postman or a gas meter reader, taking a route which doesn't involve doubling back can mean saving valuable time.

Efficiency is everything in the modern world, and managers use Euler to help them to cut corners. One nice example comes from Israel, where one branch of the main electricity company wanted to improve the efficiency of their meter readers. The job of reading

Pencil and paper challenge

This puzzle has always been popular with children. On the right is a picture of a farm gate. Is it possible to draw it without lifting your pencil and without doubling back over any line?

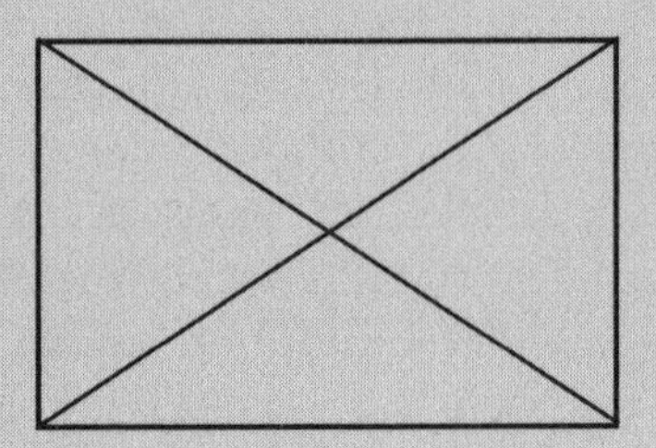

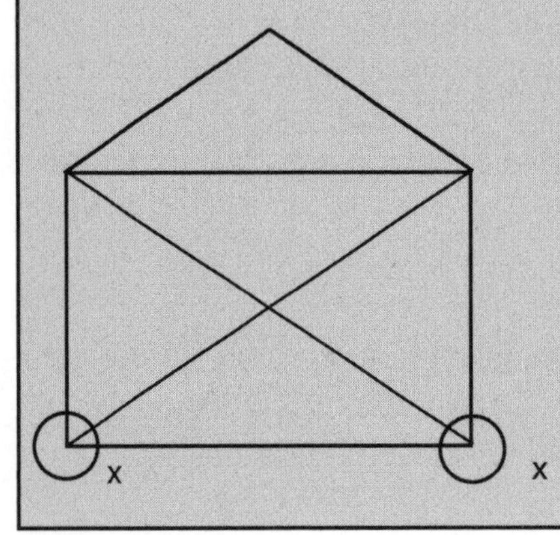

Now that you know Euler's rule you can prove that it is impossible to draw the figure, because there are four odd nodes. However, change the puzzle to the shape on the left and it becomes possible, provided you start at one of the two odd nodes labelled X.

meters in one district required 24 people, each responsible for a section of the whole network of streets. Management set about finding a way of trimming the number of people they needed to use.

The researchers into this problem adjusted the street plan for each meter reader by converting as many odd nodes as possible into even nodes. The result was a much more efficient set of routes, which led to an astonishing 40 per cent reduction in the time needed to cover the entire neighbourhood. Only 15 people were now required to read meters, no doubt leaving the other nine to curse the day that Euler ever discovered Königsberg.

Perhaps without realising it, the organisers of any sort of guided tour are also interested in Euler circuits. On a tour around a busy town, no guide wants to lead a pack of tourists back along a street that they have already visited. The problem is even more acute in a stately home whose corridors are too narrow to allow queues of people to move in both directions. Most house tours are strictly one way! This has a bearing on which doors the guides use, and which ones remain shut.

The problem becomes even more complicated for street cleaning lorries which brush the gutters on the two sides of the road separately, and therefore want to pass down each road exactly twice. This is particularly tough if some of the roads are one way streets. But all of these problems can be analysed using advanced forms of Euler rules.

Travelling salesmen

Street sweepers, postmen and tour guides are trying to avoid taking the same path twice, which means they are looking for Euler circuits. There is also a subtly different form of route which is known as a Hamiltonian circuit. A Hamiltonian circuit is a tour that visits each node exactly once but not necessarily using every path.

Here is an example. Suppose Mark sells textile machines, and he has three potential customers in his county. Today he wants to pay them all a visit. Here are their locations.

What is the shortest route for Mark to visit all his customers?

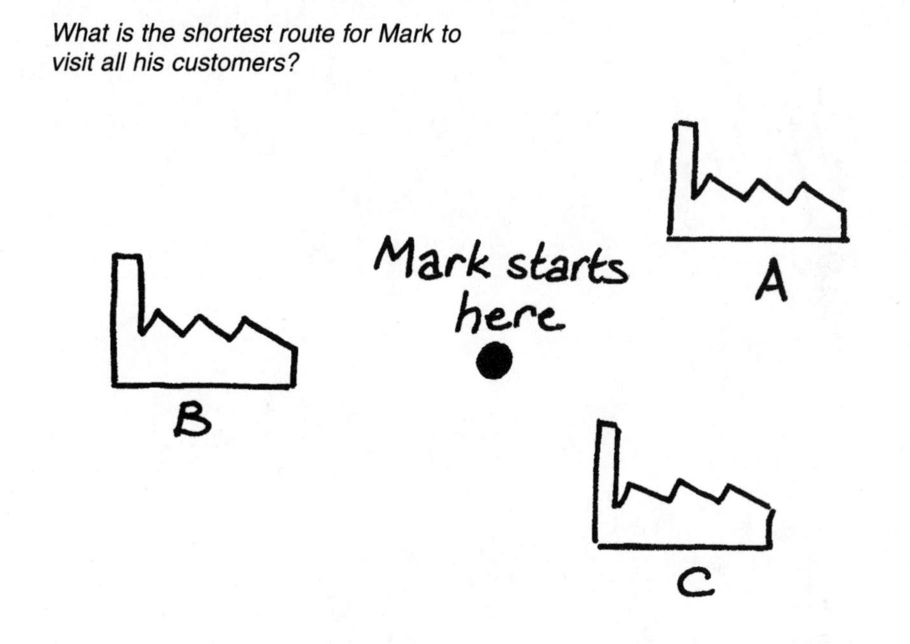

Mark has a choice. He can go ABC, ACB, BAC, BCA, CAB or CBA. In fact these are all Hamiltonian circuits, because Mark is visiting each of the 'nodes' of this network exactly once. Mark has a particular problem, however. He wants to know which of these alternatives is the shortest route, so that he can reduce his mileage and spend as long as possible with his customers.[1]

With three visits, there are six possible Hamiltonian circuits available to Mark. Six is sufficiently small that he could quickly add up the distance between each destination for each circuit and so establish the shortest route.

1 The most general problem is to visit each destination exactly once, finishing anywhere. Mark's case is special because he wants to get back home when he has completed the circuit, so there is one final leg of the journey to be added to the journeys between customers.

Mazes

Mazes and labyrinths have been around since the time of the ancient Greeks, and possibly longer. Today they are more popular than ever. The most famous in England is the Hampton Court maze which dates back to the late 17th century. A picture of the maze is shown below.

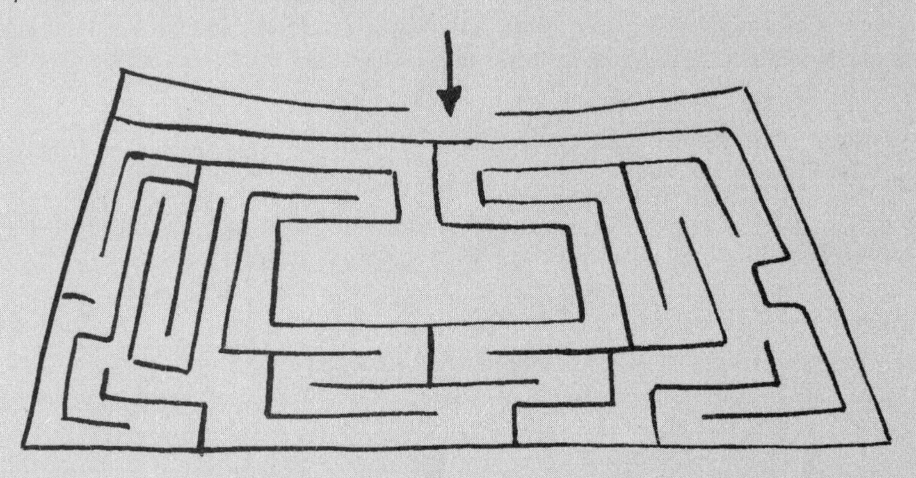

Mazes fall into two categories, 'simply connected' and 'multiply connected'. Hampton Court is simply connected. This means that it can be solved by placing one hand against one wall (choose right or left) and walking with that hand never losing contact with the wall. This doesn't guarantee the shortest route to the centre but it will always get you there eventually.

Multiply connected mazes have sections which are isolated from each other – that is, no wall joins the sections together. This means that the simple hand method doesn't work. You enter and leave the maze without reaching the centre. There is a general method for solving a multiply connected maze which has been known since the late 19th century, but it is too long to describe here (and anyway, isn't it slightly spoiling the fun?).

The quickest way to find the number of circuits is to count the number of nodes, in this case three, and multiply the descending series 3 x 2 x 1, otherwise known as 3! (or 3 factorial). If there were four customers, there would be 4 x 3 x 2 x 1, or 24 different circuits.

An exclamation mark is appropriate as the symbol for factorials, because by the time the number of customers being visited has risen to just ten, the number of different circuits available is 10! or 3,628,800. Checking which is the shortest route would be impossible without a computer. And the number of circuits escalates at a phenomenal rate. With only 20 customers, the number is so large (more than 10 quadrillion) that it is beyond even a normal computer to go through every option. For a parcel delivery needing to deliver to 60 customers, the choice of routes is astronomical.

So unlike Euler's circuits, the simple sounding problem of finding the minimum distance round a Hamiltonian circuit is in fact very hard to solve. It's all because factorial calculations for even small numbers become so huge. In fact mathematicians have still not found a general solution to the *Travelling Salesman problem* (as Mark's problem is popularly known) which guarantees to find the shortest route through a series of destinations.

This seems like a blow not only for salesmen, but also for breweries who deliver beer to pubs, doctors who go out to visit patients and indeed any member of the public who goes out to do some shopping. Almost everyone wastes a bit of petrol or a bit of time because the optimal solution is often difficult to find.

Fortunately it is not all gloom and doom. There are a number of techniques which enable the traveller to come up with a route which is likely to be close to the optimum. One of them is native human common sense – a circuit around ten destinations which is picked out by eye is likely to be within 20 per cent of the optimum distance.

To be sure of greater accuracy, a computer is needed. There are many techniques available to a computer programmer, although none of them is particularly easy to describe. Perhaps the simplest is based on what is often known as a greedy algorithm. Take for example the network on page 21.

The computer finds the two nodes which are closest to each other (D and E in this case) and puts them together on the route. Now it finds the next closest pair (A and D) and links them. If at any point it finds that the closest pair create a closed loop, as indeed linking A and E would do above, the computer drops this option and looks for the next closest pairing (A and B). It continues to do this until each node is linked to two others in a complete circuit. This technique is likely to deliver a result that is within 10 per cent of the smallest possible distance for most networks. But it isn't guaranteed to find the shortest route.

Much better techniques have been devised which guarantee that 98 per cent of the time

Stirling's incredible approximation

In the 18th century, James Stirling produced a frightening looking formula for estimating N! (N factorial).

It is:

$$\sqrt{2\pi}\, N^{(N+\frac{1}{2})} e^{-N}$$

There are two interesting points about the formula. The first is that it is astonishingly accurate – well within 1% of the right answer for factorials above 10. The second is that for no obvious reason, it features those two great numbers π and 'e' (see page 130).

Pascal's triangle and the Manhattan cabs

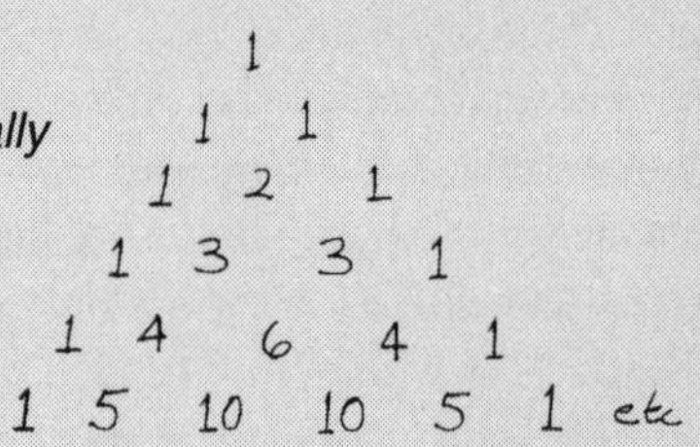

Pascal's triangle is a pretty pattern of numbers usually taught early in secondary school. The triangle is shown on the right:
To work out a number in the triangle, simply add the two above it (except the end numbers which are always 1).
Pascal's Triangle makes appearances in many real life situations, including the streets of Manhattan. In a street system where the roads form a grid, a cab driver has a choice of routes which will reach a destination in the minimum distance. For example in the 3x1 grid on the left the cab wants to travel from A to B. It has four possible routes, each of which is the same length.

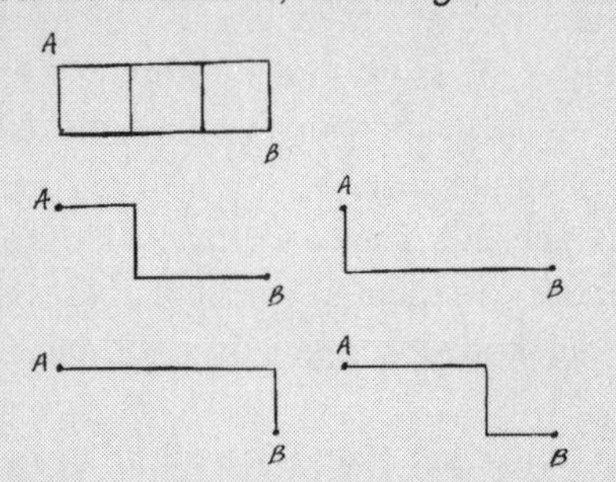

If the route is across a grid that is 2 x 2, the cab can take six routes, each of them exactly four streets long.

In fact the number of minimum-distance routes available is always a number from Pascal's triangle, and the link becomes clear if you look at the routes available for a person walking from point A (where Avenue of Americas meets 35th Street) to any other junction.
Notice how in this case the choice of routes around rectangles leads to Pascal's triangle. In Chapter 1, the choice of routes around hexagons led to Fibonacci numbers.

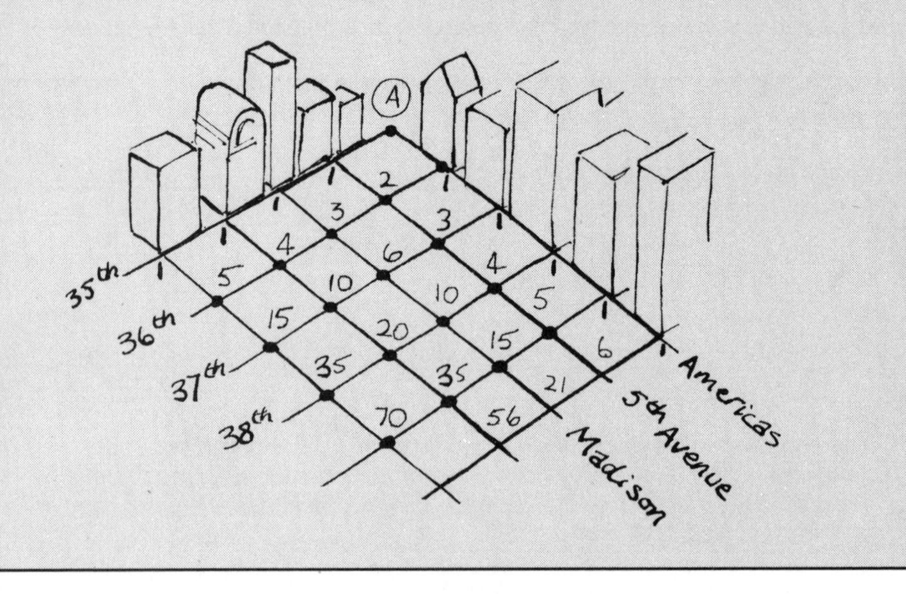

Finding the shortest route linking A, B, C, D and E

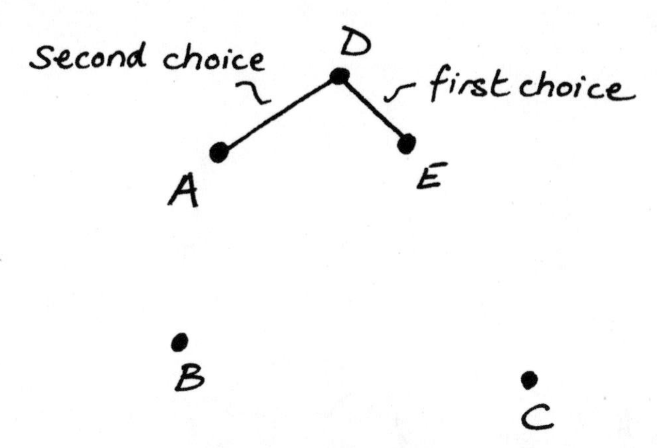

the computer will find the shortest route New technology is appearing all the time, and there are now hopes that 'biological computers', using DNA of all things, might provide an extremely efficient way of solving network problems. However, mathematicians like the purity of being able to solve a problem perfectly, regardless of any practical applications. That is why the Travelling Salesman problem has been such a challenge.

3

HOW MANY PEOPLE WATCH *FRIENDS*?

Most public statistics come from surveys, but how reliable are they?

According to the official figures published, 24 million people tuned in to watch *Friends* on 15 October 1998. That is about one tenth of the US population, a vast number of people.

But hang on a second: how do they know? Did anyone come and ask *you* whether you were watching it? Are there spies peeking in through everyone's windows? Is there some device at NBC for detecting how many TV sets are sucking their signal through the ether? Thankfully, Big Brother is not watching you. NBC knew how many people watched the program thanks to mathematics. The number 24 million was produced by the mathematics of sampling.

Sampling is the science of estimating how many people are doing something by asking just a few of them. Strictly speaking, sampling cannot guarantee to produce the true answer. If NBC wanted to know *exactly* how many people were watching *Friends* on a particular evening they would have no alternative but to monitor every household in the country. However, the cost of doing this would far exceed their budgets so it is not a serious option. But then, who needs to have the precise answer? If the number watching was really 23 and not 24 million, the program would still be shown. Many, if not most, of the statistics that we deal with every day don't have to be correct to a high level of precision. Done properly, small samples can produce remarkably accurate estimates.

Taking samples is big business. There are thousands of market research companies finding out what we eat, what we watch, where we travel, and what we think about it all. So how does sampling work?

Instead of asking 250 million people, you can get extremely close to the right answer by asking only a small fraction of that number. At least, the answer will be extremely close as long as the sample is large enough and the sample is made of a suitable cross-section of the population (to avoid bias).
Above all, however, the respondents must be telling the truth . . . .

The mathematics of lying

In most surveys, the respondents have no particular reason to lie. If somebody is asked if they bought a tin of baked beans in the last week, their response will probably be honest, even if it might be inaccurate – the memory can, after all, play strange games.

However, in other situations, lying is an issue which pollsters ignore at their peril. Always be suspicious of the answer anybody gives if they are asked for information about their income, sexual activity or, as it has emerged in recent years, politics. Questions on voting were famously the cause of embarrassment to a number of market research agencies in the 1992 General Election. At the start of the election broadcast, the results of opinion polls during the lead up to Election Day were used to forecast with some confidence that Labour would win the election, but with a hung parliament.

Little did they know, however, that the survey had overlooked a crucial distortion to the figures. It appears that while Labour voters had been happy to reveal who they had supported, many Conservatives were less willing to admit the truth. Tories felt that admitting to voting

When Truman beat the polls

One of the most famous cases in which a survey went embarrassingly wrong was in 1948 when Gallup's poll prior to the election showed that Tom Dewey would beat Harry Truman comfortably. Almost every political forecaster was shocked when Truman went on to win the election by over 2 million votes.

	Rep	Dem	Ind
NOP/NBC	40%	36%	18%
Actual Result	43%	35%	18%

Various explanations have been given for this. One is that voters changed their minds in the two weeks between the final poll and the election. But another plausible explanation is that what people said they would vote and what they actually marked on their voting slip were different. Whether you class this as lying or simply wishful thinking, it is certainly a known effect. It applied in the UK in 1992, when every opinion poll underestimated the Conservative vote by at least 4.5 percent. The final result was even outside the potential margin of error supplied by the pollsters to cover their backs!

Tory might make them sound selfish or greedy. Some Tory voters said they would vote Labour, and others simply refused to answer the pollsters. All of this meant that the reported figures were wrong, so wrong in fact that when the final result of the election was known, not only had the Tories won the election but they had won by a clear majority.

Not all lies are intended to deceive others however. Sometimes people lie in order to deceive themselves because admitting the truth to oneself can be painful. This can apply in television viewing, for example. A member of the public who discovers he watched 35 hours of TV last week may not be prepared to admit to himself that he is *such* a couch potato, and so he ticks the box labelled 21–30 hours.

One amusing example of where researchers could identify mathematically that the answers they were receiving were not truthful was a survey many years ago on the sexual habits of men and women. As part of this survey, the respondents were asked how many different members of the opposite sex they had ever slept with. The average answer for males was 3.7. The average answer for females was 1.9. Now since this survey was taken over a large and representative sample, the answers to the two questions should have been the same. After all, if a man sleeps with a woman, then the woman is also sleeping with the man, and so one is added to each tally. The conclusion that the researchers came to was that men tend to exaggerate the number of sleeping partners that they have had, while women prefer to trim the figures a little ...[1]

Statisticians need to devise suitable techniques to identify, and then eliminate, the distortion caused by lying.

Some mathematical tricks have been devised in order to help researchers obtain answers to awkward questions. During the Vietnam war, the American authorities needed to know how many of the troops were taking drugs. Drug-taking was rumoured to be high, and it was important to establish whether this was true. However, no soldier with any sense wanted to admit to taking drugs, a criminal offence. So how could the researchers get to the truth? They used a technique similar to what follows.

The researcher has a bag which contains three pieces of card, which he shows to the soldier. The three cards are:

1 This is not the only possible interpretation – for example, it could be that *both* parties exaggerate or underestimate. Another interpretation is that women find the experience more forgettable.

The soldier puts his hand into the bag and removes a card at random without showing it to the researcher. He then ticks Yes or No on the researcher's answer sheet.

If he ticks Yes, it is either because he picked the card with the black triangle on it, or because he picked the drugs question and he is admitting that he has taken drugs. The researcher cannot know which it is, and so the soldier cannot be incriminated, which means he should be more inclined to be honest.

Now here is the clever part. Suppose that the researcher interviews 1,200 soldiers in this way, and that at the end of the survey 560 have answered yes to the question on their card. On average, 400 of them picked out the card with the triangle on it, 400 selected the card without the triangle and 400 picked the drugs card. This means that of the 560 yes's, about 400 were responses to the triangle question, which leaves 160 which were drug answers. The best estimate, therefore, is that 160 out of 400 soldiers took drugs, or 40 per cent.

This is a simplification of what a market researcher would do, and we have made up the numbers here. Nevertheless, this type of survey was conducted on the US soldiers and it emerged that many US soldiers did indeed take illegal drugs during the war.

Have enough people been asked?

'In a recent test, 80 per cent of cat owners said that their cats preferred Furry Paws biscuits.' It sounds impressive, and if you are a cat owner, you might be inclined to give the product a try if you see it on the shelf. But if you now discover that this 'test' actually involved only ten cat owners, you ought to be a lot less impressed.

It is stretching belief to think that in every group of ten owners there will be exactly eight fans of Furry Paws. In fact if the researchers repeated this test, the results would keep changing. They might get the results 20 per cent, 50 per cent, 30 per cent, 0 per cent, 80 per cent. The last result would allow them to truthfully say that 'in one recent test 80 per cent preferred Furry Paws'.

It should come as no surprise that the more people that are surveyed, the more likely it is that the answer is close to the correct one. A survey of 100 people should be more accurate than one of 10, and 1,000 will be more accurate still. Taken to its limit, if the entire population is interviewed then the result is certain to be exactly right.

How large a sample do you need to take before you have enough? That depends on what you mean by enough, and it also depends on what you are testing for. In most everyday surveys, like opinion polls or tests for how many people have seen an advertisement, asking 1,000 people is usually sufficient to give a result accurate to within 5%.

However, there are exceptions. In the 1930s, the US authorities wanted to check the effectiveness of a polio vaccine, so 450 children were innoculated with the vaccine. 680 who had not been vaccinated (and who came from the same background as the test group) were monitored as a control group. Soon afterwards there was a serious outbreak of polio. None of the 450 who had been vaccinated caught the disease. Nor did any of the 680 who were unprotected. As a result, the experiment proved absolutely nothing. Even during a serious outbreak, the infection rate of polio is so low that the researchers would have needed a sample of thousands before it became likely that the control group would include polio cases, so that the experiment would have meaningful results.

Statisticians have a precise way of stating how confident they are in a result of a survey. Suppose in the cat food survey, 80 per cent say their cat prefers Furry Paws. The correct statistical way to present this result is as a range rather than as a precise figure. If 1,000 cat owners had been surveyed, then the statistician would say:

"The true figure is in the range 77 per cent to 83 per cent with a 95 per cent confidence".

This statement is easy to misinterpret. What the words in the quote mean is: "We are likely to be right by stating that the true value is somewhere between 77 per cent and 83 per cent – but there is a one in twenty chance that the answer is outside that range."

Telescopes and sampling errors

The mathematician Gauss (1777-1855) was also a keen astronomer. He acquired a new telescope, and decided to use it to produce a more accurate calculation of the diameter of the moon. To his surprise, he discovered that every time he took a measurement, his answer was slightly different. He plotted the results and found that they formed a bell shaped curve, with most results close to the central average but the occasional one quite inaccurate.

Gauss quickly realised that any measurement he took was a 'sample' prone to error but which could be used as an estimate of the correct answer. The more

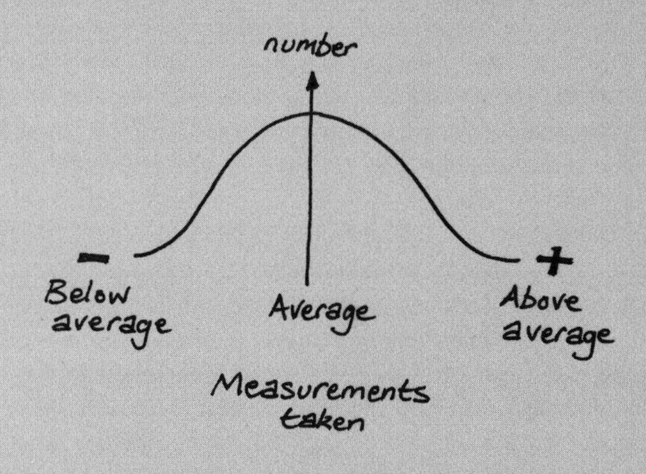

readings he took, the closer the average would be to the correct reading. He established that errors in readings belonged to a curve whose complex formula also includes π and e. There they are again!

The public has grown a lot wiser to some of the tricks of using small samples to 'prove' a result, but it still goes on. Management consultancy firms regularly conduct surveys on trends in current business, and put out press releases making claims such as '70 per cent of companies think exports are the key to success'.

Given the small samples they take in these surveys, it would be more honest if they said 'We are confident that between 50 per cent and 90 per cent of companies think exports are the key to success'. But of course no newspaper would carry a story as boring as that.

Has the average person been sampled?

The survey has been conducted. A large sample has been taken, and clever techniques have been used to ensure that the answers received have been truthful. Unfortunately this is still not enough to be certain that the results of the survey are accurate. The sample taken has to be *representative* of the whole population too.

One of the large market research companies was commissioned to test out the public's reaction to a new product: tins of baked beans with pork sausages. They selected a district of London which was conveniently located, had a representative income level and a mixed age-group. It was unbiased in every way except one. The district happened to be Golders Green. Golders Green has a very high Jewish population and a correspondingly low appreciation of pork sausages.

It doesn't matter how large the sample is, if it is biased then that bias will never disappear no matter how many people are interviewed. One of the great skills of market research is finding a sample that represents the population as a whole.

One popular way of picking a random sample is by using a telephone directory and picking out every 100th person in the list. This is a cheap way of sampling and it would be perfectly all right, for example, to use a telephone survey to find out which brand of cereal is most popular among the general public. However, it would be much less advisable to use such a survey to find out what jobs people have. Who is more likely to answer the telephone at a private address, a full time mother or a city lawyer? City lawyers' hours are probably so long that your chance of finding them at home is almost zero. And there are some professions which are seriously under-represented in the phone book. To take an extreme example, how many television presenters put their names in the directory?

Bias can crop up in all sorts of unexpected areas. A nurse often takes a patient's pulse rate for twenty seconds and scales this up to give the rate over a minute. The nurse is in fact taking a sample, and it may well be that the sample was not representative of the patient's normal state. An attractive female nurse clutching the wrist of a healthy young male can produce an extremely distorted pulse rate, especially for the first twenty seconds. If the patient is of a nervous disposition and has just been told there may be something wrong with him, this may also lead to a biased result.

Another survey that is vulnerable to bias is the one used to create the pop music charts. Perhaps you weren't even aware that the record charts are the product of a sample. Instead of monitoring the sales of a CD in every store in the country, the chart compilers nominate a number of stores to become part of their survey. The number of discs sold in these stores is added up, and the national figure is produced by scaling up the results from the sample.

Not surprisingly, the chart stores are sworn to secrecy, because if the recording companies knew which stores were used for the charts, they would immediately send out their staff to purchase discs from those stores. This would escalate their record up the charts, thus generating more coverage, and from more coverage come genuine sales. Hype is almost everything in the pop world.

This is why (it is alleged) there is a big espionage game played by the record companies to try to find out which shops are used to compile the charts. One recording company used a particularly clever if devious trick which was to pretend to be conducting market research on record shops.

They asked, 'Are you happy with the way that the chart compilers currently obtain their information from the shops?'. If the shop had no opinion or didn't understand the question, it meant that they were not involved in the charts. If they did express an opinion, good or bad, or said 'we aren't allowed to comment' it demonstrated that they knew something about what was involved and were almost certainly a chart store. In answering one question the stores were inadvertently answering another more important one!

4

WHY DO CLEVER PEOPLE GET THINGS WRONG?

Sometimes experience and intelligence can be a disadvantage.

It was a beautiful sunny morning, so the Zingerman family decided to go on a day trip to Brighton. Unfortunately lots of other people had had the same idea, and with the odd hold-up and slow road, the Zingermans' drive to Brighton worked out at an average speed of 30mph. On the return journey that evening the traffic was much worse and the Zingermans only managed an average of 20mph.

What was their average speed over the whole journey?

Adding the two speeds and dividing by 2 gives the answer of 25mph. The calculation couldn't be simpler, and that is the answer that most people come up with. Unfortunately it is wrong.

The average speed over the whole journey was in fact 24mph, and this is true regardless of whether the Zingermans lived in Bognor Regis or Birmingham.

If the answer of 24 surprises you, then you have just experienced one of the ways in which the human mind can be duped when it is solving problems. Just because a calculation looks simple and familiar doesn't mean that there isn't a surprise lurking around the corner.

The way to find an average speed is to take the *total* distance and divide by the time taken. In this case we don't know the distance, but it turns out not to matter since the answer is the same for any distance. Suppose the Zingermans' journey to Brighton was 60 miles there and 60 miles back. The journey of 60 miles to Brighton was at 30mph and so it took two hours, while the return journey was at 20mph and took three hours. This means that the average speed of the whole journey was 120 miles in five hours, or 24 miles per hour.

This average speed is known as the *harmonic mean*. It is very close to the simple mean (add up and divide by two) as long as the two speeds are relatively close together. When the British Thrust team broke the land speed record and the sound barrier in 1997, their first run was at 759 mph, and their second was at 767mph. Whichever method of averaging is used, the answer comes out at about 763mph.

This was not the case in an earlier speed record when Donald Campbell raced up Coniston Water at very high speed (at something around 300mph) but due to technical problems he limped back at more like 30mph. The average speed was published as 165mph, but the 'harmonic mean' that should have been published was about 55mph.

> ***The slow journey home!***
>
> *Speeds cannot be averaged by adding two numbers and dividing by two. This can be demonstrated with an extreme example.*
> *Suppose the Zingerman family travelled to Brighton at 30mph and that their overall average speed there and back was 15mph. What was their speed on the return journey?*
> *It is tempting to say it must have been 0mph, since (30 + 0)/2 = 15. But if they travelled at 0 mph they would never have left Brighton!*
>
> *The correct answer in this case is that they returned at 10mph to give an average of 15mph.*

Drug testers get egg on their faces

Another surprise lies in the misuse of percentages.

Medical researchers are testing a new drug called Problezene, which is claimed to improve human intelligence. Dr Smith is the first to carry out the tests on a group of his patients. As a good scientist, he decides to give some of the patients the real Problezene tablet, and the rest a 'placebo' (a tablet which has no drug in it). His results are as follows:

Dr Smith's results	Test	Successes	Average
Drug	100	66	66%
Placebo	40	24	60%

Dr Smith's results are promising. His test has successfully confirmed that Problezene is more effective than the placebo – 66 per cent of Problezene patients showed improved performance in the intelligence test compared with 60 per cent of the placebo patients.

However, because the results were quite marginal, Dr Jones decided to repeat the experiment with a larger group of patients. The results are encouraging. He confirms Dr Smith's result, as once again Problezene patients out-performed the placebo patients.

Dr Jones's results	Test	Successes	Average
Drug	200	180	90%
Placebo	500	430	86%

Excited by these findings, the two researchers decide to combine their data and publish the results, but they are embarrassed to discover a most unexpected outcome. Even though

Combined results	Test	Successes	Average
Drug	300	246	82%
Placebo	540	454	84%

Problezene was more successful than placebo in both of the tests, when the tests are combined, the placebo patients turn out to have been more successful than the Problezene. This is such a surprising result to some people that they regard it as the mathematical equivalent of an optical illusion. Where is the typing error? The only error is in the logic which assumes that percentages can be combined in the same way as simple numbers. Percentages cannot be added and averaged, just as speeds cannot be averaged.

Ring around the rodeo

Every year, Texas businessman Wild Bill Mahoney holds a rodeo for the cowboys on his ranch. In order to allow enough room for the rambunctious activities, the fence for the corral must be big enough to ensure that the center of the corral is 30 metres from the fence all the way around. Unfortunately, somehow this year Wild Bill has lost 6 metres of his fence, which means that space will be tighter this year for the bucking broncos—but by exactly how much? (Does your gut feeling say a centimetre? A metre? More?)

The answer is that the corral fence will be one metre closer to the center all round, or exactly 29 metres from the center in any direction. That answer doesn't surprise many people.

Now it just so happens that out there in the cosmos, the gods also like to hold a giant rodeo. Their intergalactic corral is vast, a million miles across, and the gods also like to put a fence around the event to make sure the cosmic stallions cannot escape. By an astonishing coincidence, the gods have also mislaid 6 metres of fence this year, so again the space for activities will be tighter, but by how much this time?

In this case, the gut reaction might be that the fence will only be a fraction of a millimetre closer in. After all, there is so much area to cover that six metres is bound to get spread around the edge very thinly. It may come as a surprise, therefore, to learn that the gods' corral is also one metre smaller all the way around this year.

How come? Here is the calculation. The perimeter of each rodeo corral is the length of picket fence. The radius is the distance from the centre of the corral to the boundary.

$$\text{Perimeter of circle} = 2\pi \times \text{radius}$$

In both examples, this year's perimeter (which we will call New P) is six metres less than last year's (Old P), and what we want to know is how much the radius R of the corral has changed, (Old R – New R).

$$\text{Old } P = 2\pi \times \text{Old } R$$
$$\text{New } P = 2\pi \times \text{New } R$$

We know that Old P – New P is six metres, and so:

$$6 = 2\pi \times \text{Old } R - 2\pi \times \text{New } R$$
$$= 2\pi \times (\text{Old } R - \text{New } R)$$

So Old R – New R (the change in the distance to the boundary) is $6/2\pi$, which, using π as 3.14, gives an answer of about about one metre. The length of the corral fence is irrelevant to this calculation! This can come as quite a surprise because intuition tells us otherwise.

Even the greats make mistakes

In 1935 a Frenchman published a book called ***Erreurs de Mathematiciens des Origines a Nos Jours*** *which listed errors by 355 mathematicians including Fermat, Euler and Newton. Sometimes mistakes are almost inevitable because of the complexity of the problem. When Andrew Wiles first published his proof to Fermat's Last Theorem in 1993, it contained a crucial error, but one which only the top mathematicians were able to understand, let alone spot. Other mistakes are forgivable because of the lack of knowledge at the time. Isaac Newton believed in alchemy and was convinced that lead could be turned into gold. It seems strange now, but was less so in his time, given how little was known about chemical elements in the 17th century. Some mistakes are less easy to excuse, however. The absent-minded professor can often be extremely prone to elementary mistakes of arithmetic, and can be very gullible to the simplest of tricks. One reason for this is that a person of very high intelligence often looks for complexity in a problem when there is none there.*

The whisky mix up

After dinner, Henry Bufton calls for his butler who brings him his usual half glass of whisky and a glass of water. Henry pours a small amount of water into the whisky. Realising that his whisky glass is now too full, he gently pours back some of the whisky–water mix into the water glass until his whisky glass is half full again.

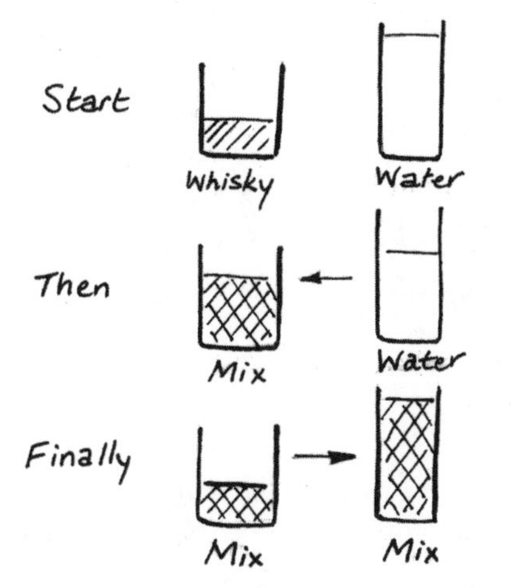

Is there more water in the whisky than whisky in the water

Henry poured pure water into his whisky and then a mixture of whisky and water back into the water. The question is this. Is there now more water in the whisky glass than there is whisky in the water glass?

A common answer is to say that there is more water in the whisky glass. After all, it was a pure sample of water that went into the whisky, but only diluted whisky went in the other direction. Both glasses now contain exactly the same amount of liquid as they did at the start. By now, however, you are beginning to suspect that it cannot be this simple. The correct answer is that the amount of whisky and water transferred are the same.

This answer is often disputed. The best way to demonstrate why it is true is to imagine that instead of glasses of liquid, there are two buckets of tennis balls. To start with, one bucket contains 100 green balls. This represents the water. The other contains 20 white balls. This represents the whisky.

Take any number of greens – we'll use 10 – and transfer them to the white bucket:

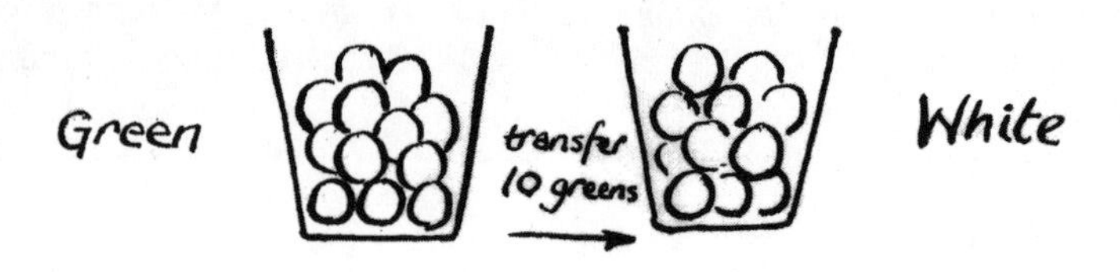

After this transfer, there are 90 greens in one bucket, and 20 whites and 10 greens in the other.

Now transfer 10 balls back, but this time a mixture. Suppose it is 8 whites and 2 greens:

After this second transfer, there are 92 greens and 8 whites in one bucket, and 12 whites and 8 greens in the other. The two buckets contain the same number of balls as they did at the start, but eight greens (the 'water') have swapped buckets with eight whites (the 'whisky'). It doesn't matter what mixture is taken back, the number of green and white balls swapping buckets are always the same.

Convinced? If not, the best way to prove it to yourself is to experiment with some real balls. Better still, try some real whisky.

A quick sum

Some of these problems can be hard to get your brain around, which can help to explain why mistakes are made so often. Not so with the final one. Here is an extremely simple addition sum. Cover up the whole sum with your hand, then reveal one number at a time as you add up the numbers in your head:

```
1000
  40
1000
  30
1000
  20
1000
  10
____
```

What answer did you get?

If your first answer was 5000 then look again, because it is not right. The correct answer to the sum is 4100. When this problem is sprung on them without warning, most adults make the same mistake. After reaching 4,090 the brain anticipates that the answer will be rounded up. It assumes from previous experience that the rounded number will be a nice easy one, so it says 5,000 without thinking.

Sometimes a brain can be too clever for its own good.

5

WHAT'S THE BEST BET?

Lotteries, horses and casinos
all offer the chance of a big prize.

Most areas of mathematics developed from uplifting and noble research. However, there was one notable exception. Probability theory, one of the most important areas of all, had its origins in vice.

Galileo is often remembered as the Italian who confirmed that the earth moved around the sun, and was then forced by the church to retract this heresy. Suggesting that the Bible might have got it wrong wasn't his only sin, however. Galileo also advised one of his patrons on how to bet in a dice game, and gambling was an activity disapproved of by both society and the church.

Ten years after Galileo's death, Pascal and Fermat developed the subject properly, but they too did so because wealthy noblemen wanted to increase their winnings.

What Pascal and Fermat asked themselves was this: 'When should I gamble, and when should I stop?' Those are also the questions behind this chapter.

Coins and dice

The simplest bet of all is on the toss of a coin. I will give you $10 if it's heads, you give me $10 if it's tails. This is so obviously a 'fair' bet that the maths is taken for granted. Nonetheless, let's go through the basic mathematics which can be applied to more complicated bets later on.

Everyone knows that the chance of a head on the toss of a coin is 'Fifty-fifty'. This is the language of odds with which people are most comfortable. There are, however, at least six different ways of describing a probability, all of which mean exactly the same thing. The chance of tossing a head on an unbiased coin can be expressed as any of the following:

- Fifty-fifty
- 1 in 2
- 1/2 – mathematicians often quote probabilities as fractions
- 0.5 – mathematicians also like using decimals
- 50 per cent – percentages are favoured by weather forecasters for some reason
- Evens – this is the language used by bookmakers (see the box opposite).

What they all mean is that if you toss an ordinary coin 100 times, you expect it to come up heads 50 times. Sometimes more, sometimes fewer, but on average 50.

To work out how much you expect to make from a bet, you need to look at how much you stand to win or lose for each possible outcome, and the chance of each outcome happening.

In the example of heads and tails with $10 at stake, suppose that you call heads. Here are the possible outcomes:

Result	Chance of it happening (P)	How much you will win (W)	PxW
Heads	½	$ 10	$ 5
Tails	½	−$10	−$5

Is this bet worth taking? This is where the final column (PxW) comes in useful. If you add up the column, it gives you the expected value of the bet. In this case the value is $0, which means that on average you should end up no better off than when you started. But at least it also means you should end up no worse off either, which makes this better than most bets available to you on the market!

Now for a slightly more complicated bet. Harold has a fair die, numbered one to six. If he rolls the die and it comes up six, he will pay you $24. If it comes up anything but six, you pay him $6. Is this a good bet for you?

To assess this bet, you need to know the likelihood of throwing a six. The probability of throwing a six is 1 in 6 or 0.16666 or '5 to 1 against' in bookmakers' language. The chance of not throwing a six is 5/6. Here are the two outcomes.

Result	Chance of it happening (P)	How much you will win (W)	PxW
Throw 6	1/6	$ 24	$ 4
Don't throw 6	5/6	−$6	−$5

↑ notice that column P adds up to 1 again

Bookmakers' language

Bookmakers have their own language for expressing odds. The chance of rolling a three on an ordinary die is 1 in 6, since the die has six sides. However, a bookmaker who was not out to make a profit would quote this as 5 to 1 against (that is, five times in six the gambler who bets on a three will lose). The chance of turning up the Ace of Spades in a pack of cards is 1 in 52, but to a bookmaker it is 51 to 1 against. Notice how the bookmaker always puts the larger number first.

If the odds of winning or losing are exactly equal, the bookmaker calls this Evens.

And if the odds are such that you are more likely to win than lose, the bookie replaces the word 'against' with 'on'. The chance of rolling a number bigger than two on a die is 4 in 6, which to a bookie is 4 to 2 on, except that he would reduce this to its simplest form, which is 2 to 1 on.

The final column now adds up to minus $1, which means that on average for each roll of the die you stand to lose $1. If you rolled the die one million times, you would expect to lose $1 million.

And as far as the gambling industry is concerned, that is how it should be. The whole point of gambling is to give you the prospect of making a fantastic return on your investment of a bet in a single turn, while ensuring that in the long run the organiser will make a profit.

Lottery

For the UK National Lottery, there is no need to work out what your expected return will be: they have made that explicit. For every £1 paid into the lottery, 50p goes in prize money, and the rest goes in tax, good causes and administration. That means that your expected return on each bet is a loss of 50 pence. So every time you don't enter the lottery you could say to yourself 'Hey, I've just made another 50p profit!'

Of course the reason why people enter the lottery is that they feel they wouldn't miss £1 lost from their wallet, but they would certainly notice £10 million on their bank balance. Lottery entrants would argue that the excitement they get when waiting for the numbers to come up is worth at least 50 pence in fun, although perhaps they don't take into account the negative value of the depression which builds up as, week after week, the numbers don't come up. People who get a kick out of taking risks are known mathematically as risk-friendly.

The UK Lottery involves picking any combination of six numbers between 1 and 49. Six balls are drawn at random from a tub containing 49 balls. If these six balls happen to have your six numbers on, then you win the jackpot, which is typically £10 million. At least, you win a share of the jackpot. If you are lucky, nobody else selected your combination of numbers, so you get the entire jackpot to yourself. More often, two or three people pick the same combination and so share the jackpot.

Which is the best combination to pick? Rather than thinking about choosing 6 balls from 49, imagine a simple version. Suppose there were only three lottery balls, and you could choose only two of them. These are the choices you could make:

1,2
1,3
2,3

The order that the balls come out of the tub doesn't matter. If you select 1 and 3, then the balls can come out of the tub in the order 1 then 3, or 3 then 1 – you win either way. So there are three possible combinations, only one of which will win. But are they all equally likely? The answer is yes, and to prove it you can list every possible outcome of drawing balls from the tub:

1,2 1,3 2,1 2,3 3,1 3,2

There are six possible permutations for drawing balls, and each of the three combinations appears twice, making each combination equally likely. The probability of winning a share of the jackpot in this mini-lottery is therefore 1/3.

In fact you can use this logic to show that however many balls there are in a lottery, and however many numbers you are asked to pick, each combination is just as likely to win as any other. In other words, if your choice is 1,2,3,4,5,6, it is just as likely to be the winning outcome as 11, 17, 20, 31, 34, 41, even though the latter seems much more 'random'.

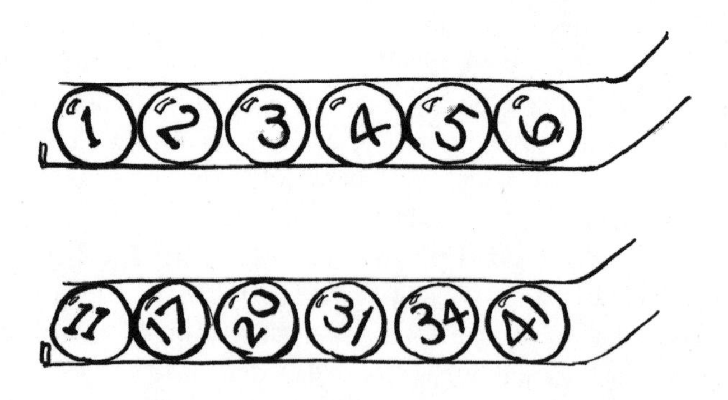

The number of combinations possible turns out to be huge: 13,983,816 to be precise.

Although you cannot influence your chance of hitting the jackpot, you can at least choose numbers which increase your expected financial gain. You do this by picking a combination of numbers that is likely to be unique, so that if you do pick the right numbers, you are more likely to have the jackpot all to yourself.

Surprisingly one of the worst combinations to pick is 1, 2, 3, 4, 5, 6 – hundreds of people pick this combination. Presumably they work on the assumption that 'nobody else would think of something this unusual'. Unfortunately a lot of people would!

Because many lottery entrants like to use 'lucky' numbers, and lucky numbers are often linked to birthdays, more people are likely to select numbers between 1 and 31 than between 32 and 49. So, if you don't want to share your jackpot with others, it is worth biasing your entry towards higher numbers. Watch out though, a lot of people know this theory now. This is an example of the only valid strategy we know of for edging out others playing the lottery, and it has been called the avoidance strategy. The basic principle is to avoid picking numbers that are deemed 'special.'

There is lots of mumbo jumbo about selecting numbers, which mathematicians would dismiss. For example, the idea that '39 hasn't appeared for three weeks so one must be due this week' is superstitious nonsense. If you toss a coin ten times and it comes up heads each time, the eleventh toss is no more likely to come up tails than the first. Indeed if anything you might expect another head, since it is beginning to look as though the coin is biased towards heads. If the number 39 comes up in the lottery every week for six weeks, it is just possible that there is a physical reason favouring the 39 ball – it may be a tiny bit heavier than the others, for example. But since lots of people would start picking 39 if it appeared frequently, your best tactic might be to avoid it!

Why should Californians buy their lottery tickets on Friday?

The answer to this is that if a Californian buys a ticket earlier than Friday, their chance of winning the lottery jackpot the following Saturday is lower than their chance of being run over by a car before they are able to claim their prize.

This rather gruesome statistic comes from the fact that the chance of winning the California lottery in any given week is roughly 1 in 18,000,000 and the chance of being run over by a car in that state during a 24-hour period is roughly the same. This isn't intended to make you scared about walking outside, incidentally. The statistic says more about how unlikely you are to win the jackpot than about how likely you are to be run over.

Horses and bookies

In the betting games described so far, the operator of the bet has made a profit by fixing it so that the expected pay-out is less than the odds are worth. But what about horse racing, where the odds are such that if a horse is 'evens' you pay the bookie $10 if it loses, but he pays you $10 if it wins. It sounds as if he is being astonishingly fair.

A simple table can be used once again to demonstrate why bookmakers usually have a smile on their faces. Suppose there is a race with three horses with the following odds.

Naughty Nibbler	Evens	(= 1/2)
Fiery Fred	2:1	(= 1/3)
Old Lag	3:1	(= 1/4)

Andrew, Bert and Charlie are three brothers who each place $1 on a different horse. One of them will win, but how will the family do overall?

Result	Odds offered (P)	How much winner gets (W)	P×W
Naughty Nibbler	1/2	$1	50¢
Fiery Fred	1/3	$2	67¢
Old Lag	1/4	$3	75¢

That's curious. According to the table, the three brothers' combined expected winnings come to $1.92, yet the two who lost each paid $1. The bookmaker expects to make an average of 8 cents profit. How does he do it?

The explanation comes from looking at the column labelled 'odds offered'. In the other betting systems we have looked at, the odds add up to exactly 1.0, but in this example they add up to 1.08 (i.e. more than one). In other words, a bookmaker makes money by *fixing the odds* so that they are artificially short, rather than fixing the prize money. If you ever discover a bookie whose odds add up to less than one, place your bets with him immediately!

There is one important difference between the odds here and those in the lottery. The chance of a combination of numbers appearing in the lottery is known exactly, but the chance of Naughty Nibbler winning the race is only a guess, albeit a well-informed one. One reason why punters bet on horses is that they feel that their own knowledge is better than that of the other punters. So while 8 to 1 may be the quoted odds, if you believe the horse really has only a 10 to 1 chance of winning, or if you know that the horse is actually suffering from food poisoning, then you should place your bet elsewhere.

Incidentally, how often do horses quoted at 2 to 1 actually win a race? It would be interesting and quite easy to find out, by examining the result of every race this year in which a 2 to 1 horse ran. Of 100 winners in these races about one in three, that is 33, should have been the 2 to 1 horses. If there were only 10 such winners, this suggests that 2 to 1 horses are not generously priced as a rule, so don't bet on them. If on the other hand there were 50 winners, it sounds as if 2 to1 horses are a good bet, although it would only remain as a good bet if you kept your findings to yourself. If they became widely known, everyone would start betting on 2 to 1 horses and the odds would get worse!

Is there a bet that you cannot lose?

It may be beginning to sound as if only a sucker could be drawn into a bet with a professional. The lottery takes 50 per cent of your money, the football pools a similar proportion and bookmakers take over 15 per cent. And all of that is ignoring the tax you might have to pay on top. Even roulette wheel operators take their cut. The wheel is numbered 1 to 36, but there is also a zero slot, and if the ball lands there then all the money goes to the casino.[1] This happens on average 1 in 37 times, giving the casino a 3 per cent cut – which is actually pretty generous to the punter compared to some other bets we have looked at.

There is one type of bet which, remarkably, seems to guarantee that you will win eventually. You can use it in any situation where the odds are 50-50 that you will win (in other words, when the odds are evens, if you will excuse the pun). It is called the martingale.

First of all, decide on how much you want to win. $10 seems like a reasonable amount, not too greedy. Place your bet of $10. If you win, congratulations – take the $10 win, and finish gambling now.

If you lose, then you need to bet again. This time, place a $20 bet. If you win, then collect the $20 and stop gambling. Your winnings are $20 for the final game minus $10 lost in the first game giving you a profit of $10 net.

If you lose, you need to double your stake again, this time to $40. And in fact the rule of the martingale bet is that every time you lose, you simply double your stake and bet again. Even if you lose fifty times, if you win the fifty-first time then you end up with a profit of, you guessed it, $10.

1 The exception is betting red/black, odd/even or high/low. For these even money bets, half the stake is returned if the ball lands on zero.

When you do eventually win, your net gain will be exactly the same as the amount that you stake in the first bet. So, if you are greedy and want to be guaranteed a win of $1,000,000, then that is what your first bet should be.

Does this sound too good to be true? Why are we telling you this when we could have made our million and be sitting on a beach in the Bahamas?

Surprise, surprise, there is a catch. Although the theory is perfectly sound, in practice the martingale bet would never work, for one simple reason. Casinos and bookmakers set upper limits on the amount that you can place on a single bet. Even if you wanted to place a $10-million bet, you would not be allowed to do it. And even if you could, which bank is going to provide you with the security that allows you to place the $20 million bet that follows if you lose?

The truth is that there is no easy, foolproof way to make a million pounds. For everybody who wins a bet there has to be somebody who loses. The betting industry thrives because there is a sucker born every minute, and fortunately for them this doesn't look as though it is about to change.

Betting on the bizarre

The bookmakers Ladbrokes are prepared to consider almost anything as a potential area for a bet, though not surprisingly it is rare to find them offering odds above 5000 to 1.

The odds of Halley's Comet colliding with Earth on its next visit are quoted at only 2500 to 1, and the odds that the United Nations will confirm the existence of alien life forms were recently shortened from 200 to 1 to just 50 to 1.

Ladbrokes won't take bets on everything, however. One man wanted to wager that his wife was going to be abducted by aliens and that at the stroke of midnight at the start of the year 2000 she would return as a teapot. The bookmakers gracefully declined to offer any odds on this event.

6

HOW DO YOU EXPLAIN A COINCIDENCE?

Coincidences aren't as surprising as you would think.

At a recent seminar on the paranormal, some adults were asked if they had ever experienced an interesting coincidence. Most of them had. One woman recounted how, the previous year she had been on holiday in Switzerland. It turned out that the family staying in the next chalet were her former neighbours from home.

More spooky was the man – call him Tony – who said that as a schoolboy he had really disliked his headmaster. One Sunday night he dreamed that his headmaster had died. The next morning he arrived at school to discover that the headmaster had indeed snuffed it over the weekend. The seminar room went rather quiet after hearing Tony's story, then somebody whistled the tune from the *Twilight Zone* and everyone burst out laughing.

It is interesting that so many people prefer to attribute coincidences to some mystical force than to a more rational explanation. This has something to do with the human psyche. People love mystery and the paranormal, hence the enormous success of programmes like *The X Files* on television. However, the belief in hidden forces is also partly to do with a general ignorance in society about the mathematics of chance.

The immediate conclusion we would all be tempted to make in the dead headmaster story is that Tony had some lethal psychic power. There are, however, a number of other perfectly rational and perhaps more plausible explanations. The first is that Tony had forgotten the precise details of the story – it wasn't a dream, but more a case of *déjà vu*. Another possible explanation is that Tony knew that the headmaster was seriously ill, which meant that it was quite likely that death would come soon and that he might dream about it. It is also not unknown for a piece of news to enter the subconscious, and for it then to suddenly appear, as if from nowhere, as a spontaneous thought. Perhaps Tony's parents heard the news of the headmaster's death and Tony overheard them talking about it while he was doing his homework.

There is, however, another possible explanation. Maybe the headmaster's death was a complete coincidence. One of the many definitions of coincidence given in the Oxford English Dictionary is: 'a remarkable concurrence of events without apparent causal connection'.

It is an event which seems so unlikely that it is worth telling a story about. But should we be that surprised? How likely is a coincidence? If coincidences cannot be predicted by mathematics alone, maybe we should start believing in the paranormal after all.

Presidential coincidence

One strange historical coincidence concerns Presidents of the USA. Three of the first five Presidents died on the same day of the year. And that date? It was none other than 4 July. Of all the dates to die on, that must surely be the most significant to any American.
This might of course be part of the explanation as to why the coincidence happened. You can imagine the early Presidents being really keen to hang on until the anniversary of Independence, a date which meant so much to them, and giving up the ghost as soon as they knew they had reached it. This is apparently what happened to Thomas Jefferson, the third US President. John Adams, the second President, actually died just a few hours after Jefferson. His final words were 'Thomas Jefferson still survives'. He was wrong.

Birthdays

The first thing to say about coincidences is that we can often be over-impressed by them. During a lesson given to a class of 30 primary school children, the subject of birthdays came up. One of the children announced, 'Sally and me have the same birthday'. This was very special for them both, and exciting to the rest of the class.

What may seem odd, however, was that this was not an unusual story. If you go into any school class, more often than not you will find at least two children with the same birthday. To most people this might seem an unlikely coincidence. After all, there are 365 days in the year, so you might expect that you needed a classroom with about 180 children in it before there was a fifty-fifty chance that there would be a coincidence of birthdays.

However, this is not the case. Surprisingly, you only require 23 children in a class for there to be more than a 50-50 chance that two of them have the same birthday. In fact because birthdays are not spread evenly across the year, if classes had only 20 children you would probably still find that there was a birthday coincidence in half of them.

How can this be? To work it out, you need to know that in order to calculate the probability of two 'independent' events happening together, you multiply the probability of each of the events together (see the box). For example, the chance of tossing two heads on a coin is 1/2 x 1/2 = 1/4, or 1 in 4. To convince yourself of this, toss two coins and see what combination comes up. Repeat this experiment one hundred times and count the number of double heads. There should be about 25. There is no guarantee of exactly 25, but if you don't get between 20 and 30 you may be tossing the coins in a peculiar way.

Like the toss of a coin, one child's birth date is independent of another child's (as long as they aren't twins!). This means that you can calculate the chance of a birthday coincidence by multiplying probabilities together in the same way as for coin tossing. But instead of calculating the chance of a coincidence, let's work out the chance that all of the children have *different* birthdays – it's actually a much simpler calculation.

Imagine first of all that you have only two children in the class. The first one's birthday is 14 June. What is the chance that the second child's birthday is different? There are 364 other dates to choose from, so the probability is 364/365 that the two children have different birthdays. Sarah now enters the room. Does she have a birthday different from the other two children? If the other two children have different birthdays, then the chance that Sarah's birthday is different again is 363/365 – there are only 363 days left which are different. Simon is next to arrive. The chance that his birthday is different again is 362/365 ... and so it continues.

As each new child enters the room, the chance of them having yet another different

How many men wear skirts?

What's the chance of tossing a head and then rolling 3 on a die? Because tossing a coin has no influence on rolling a die, the chance of tossing a head and rolling a 3 can be calculated by simply multiplying the chance of the two events together.

The chance of tossing a head is 1/2

The chance of rolling a 3 is 1/6

The chance of both together is

1/2 x 1/6 = 1/12, or a 1 in 12 chance.

This doesn't work if the two events are not independent, however. For example:

- *The chance of a person in the street being male is about 1/2*
- *The chance of a person in the street wearing a skirt is probably about 1/4*
- *However, the chance of the person in the street being a man wearing a skirt is not 1/8, because a person's gender influences their tendency to wear a skirt.*

The influence of one event on another is the basis of Bayesian statistics, an important part of probability theory.

birthday diminishes ever so slightly. The chance of the 23rd child having a different birthday from everyone else is 343/365.

And at this point, let's stop to work out what the overall probability is that each of the 23 children has a different birthday. We calculate it by multiplying the probabilities together, just as we did for the coins:

$$\text{The probability of no person having the same birthday as someone else in a room of 23 people} = \frac{364}{365} \times \frac{363}{365} \times \frac{362}{365} \times \ldots \times \frac{343}{365}$$

$$= 0{\cdot}49 \text{ or a } 49\% \text{ chance}$$

So the chance of no two children in a class of 23 having the same birthday is 49 per cent, or about half. But what is the alternative to no children having the same birthday, the other 51 per cent? It is that at least two children do have the same birthday. In other words, the chance of at least one birthday coincidence is 51 per cent with just 23 children. This result doesn't feel right to many people, but it is true. Furthermore if you don't believe it then you can test it for yourself with a visit to your local school.

But what has this got to do with bumping into a friend you haven't seen for twenty years when you only mentioned him to your wife last night. To examine such an apparent coincidence, the first thing to do is distinguish between:

- The chance of a *particular* unlikely event occurring
- The chance of *any* unlikely event occurring.

In the case of the school children, you might immediately see where these two types of coincidence are being confused. If you pick two specific children, David and Charlotte, out of a class of 23, the chance of them having the same birthday as each other is 1 in 365. But you now know that the chance of there being at least one pair of children in the class with the same birthday (though we've no idea which pair) is 1 in 2, which is massively more

Guessing game

Here is a game you can try with a group of ten people. Ask each person to write a whole number between 1 and 100. The aim of the game is to write down a number that nobody else will write down. You would think this would be easy to achieve, but even if everyone picks randomly, the chance of two picking the same number is more than one in three. And in fact, because people DON'T pick randomly (numbers above 50 are more popular for example), the odds are about 50-50 that two people will pick the same number. The more people that play, the more the chances of a coincidence escalate. With 20 people, the odds of a coincidence are about 7 to 1 on (strong enough for you to do this as a mind-reading stunt).

likely. Yet the two coincidences both *feel* about the same. It just goes to show that gut feeling can sometimes be very misleading.

In every other walk of life, there is a similarly huge difference between the chance of a specific coincidence happening such as 'within a week of reading this sentence you will meet an old schoolfriend in the street', and any old coincidence happening like 'something interesting will happen to you in the next week'. Both will grab your attention, but the latter is far more likely.

Think about how many 'events' you experience in a day. You get up, brush your teeth, have breakfast, listen to radio, get in the car, more radio, drive to work, meet lots of people, make lots of phone calls, day dream, have lunch ... it goes on and on. You experience literally hundreds of events every day. Each one gives you the opportunity to have a coincidence and most days these events go unmentioned. Let's face it, it would be pretty sad if every evening you went home and declared to your partner:

"What a day. I met a woman called Jenny Stewart but we didn't have any friends in common, everything I dreamt about last night didn't come true, and just as I was leaving my office nothing happened'. You wouldn't say this because these events are all boring. They were the coincidence opportunities that failed.

There are also *unlikely* events that happen to you that you don't classify as a coincidence. Just because something is unlikely, it doesn't necessarily make it interesting. To take an extreme example, let's pick two famous people completely at random. Queen Victoria and George Washington. Queen Victoria's birthday was 24 May and George Washington's was 22 February. Now that is incredible. The chance of Victoria's birthday being 24 May was 1 in 365, as was the chance of Washington's being 22 February. So the chance of Victoria and Washington having those exact birthdays was about 1 in 130,000. Whoever would have thought that something so unlikely as this would happen ... but wait, you don't seem very impressed.

The reason you are not impressed is because it begs the question 'so what?' Two events have coincided here, but you wouldn't call them a coincidence. Boring incidents are quickly forgotten, but coincidences grab the attention and stick in the mind.

Amazing life coincidences

But how *likely* is an interesting coincidence to take place?

Let's make a rough estimate. Suppose that a really memorable, once in a lifetime coincidence is one which has a one in a million chance of happening today, and that during any particular day there are 100 opportunities for one of these extremely unlikely coincidences to happen to you. For example, you decide on a whim to place a bet on three rank outsiders in the Grand National, and they end up first, second and third in the race. Or it is General Election day, and you have a minor road accident as you are driving back from the polling booth only to discover that the passenger in the other car is your old MP. These are 'one in a million' type coincidences. So, incidentally, is the chance of dreaming about a friend winning the lottery and it coming true within a few days.

As with the birthday example, the best way of working out the chance of such a coincidence happening is to look first at the chance of there being no such coincidence.

What is the chance that *none* of these coincidences happens to you tomorrow? The chance of a one in a million event not happening is 0.999999. We have estimated that 100 opportunities for such an event happen each day, so the chance that none of them happens is:

0.999999 x 0.999999 x 0.999999 . . . one hundred times.

This comes to roughly 0.9999, or 9,999 in 10,000. Which means that the chance that one of these coincidences *will* happen to you tomorrow is 1 in 10,000. Still very unlikely.

How does the next week look? What is the chance that on each of the next seven days, there will be no one-in-a-million coincidences? We calculate as before. The chance of having a day as boring as this one every day this week is:

0.9999 x 0.9999 x 0.9999 . . . seven times.

or roughly 0.9993. This means a 9,993 in 10,000 chance of a boring week, and 7 in 10,000 chance of a fantastic coincidence during the next week.
The chance that every week for the next year will be this dull is:

0.9993 x 0.9993 x 0.9993 . . . fifty-two times.

or 0.964, which is about 29/30. Suddenly this is beginning to get interesting. The chance that every one of the next twenty years will have no one-in-a-million coincidences for you is:

0.964 x 0.964 x 0.964 . . . twenty times.

This is equal to 0.48, or a 48 per cent chance.

According to this extremely rough and ready calculation, there is actually more than a fifty-fifty chance that in the next twenty years you *will* experience a memorable one-in-a-million coincidence. This also means that for every twenty people you know, there is a greater than 50% that one of them will have an amazing story to tell during the course of a year. So just as with the class of 23 children, the chance that one of you will have a fantastic coincidence this year is 52 per cent, more than half! Maybe life isn't so boring after all.

Of course we have made some huge assumptions here. Who knows how many 'astonishing' coincidences there are that could happen to you on any day? Maybe there are thousands of them. Some of them are actually one in a billion chances, while others may be one in a thousand. But our very rough estimate probably isn't that far out, and a 50-50 chance of something really interesting happening to you or one of your closest

Horoscopes and coincidences

Here is your horoscope for today. Do you agree with it?

At times you are extroverted, affable, sociable, while at other times you are introverted, wary and reserved. You have found it unwise to be too frank in revealing yourself to others. You pride yourself on being an independent thinker, and do not accept others' opinions without satisfactory proof. You prefer a certain amount of change and variety, and become dissatisfied when hemmed in by restrictions and limitations. You have a great deal of unused capacity which you have not turned to your advantage. You have a tendency to be critical of yourself.

If that description strikes you as uncannily accurate then you have been a victim of what is known as the Barnum Effect. This is more meaning into a situation than there really is. Bertram Forer published the first paper on the subject in 1949. Most of the above statements are 'true' for the majority of people. Those statements which are not true will tend to be ignored, while those which contain truth will be noted.
Researchers have found that if the star signs on a horoscope are removed, people are unable to identify which paragraph belongs to their own sign, but if the signs are included they will believe their own star reading to be the most accurate.

friends this year shows that you shouldn't be that surprised when you hear about someone's spooky coincidence.

Even so, that is easy to say until it happens to you. Here is a true story that was given to us:

> *A couple of years ago I was visiting a new acquaintance. Her young daughter Sarah was there with some crayons. I drew a picture of a moon for her and said 'and of course you can tell what date it is by the shape of the moon!' (I was making this up) 'and the date is.... randomly thinking of a date ....17 August'. The mother gasped. 'I knew you would say that,' she said. 'Sarah's birthday is 17 August, and so is mine, and so is my husband's'.*

It was a monumental coincidence. A coincidence as interesting and unlikely as that may only happen to a person once in a lifetime.

7

WHAT'S THE BEST VIEW OF THE STATUE OF LIBERTY?

Everyday geometries, from snooker to statues.

Try asking adults which part of the math they did at school they found the least useful. 'All of it,' is one common response. But probe more deeply, and geometry and trigonometry are two topics most likely to provoke a stifled yawn. Indeed one American friend recalls that she once asked her teacher during a lesson 'can you please explain why we are doing all this geometry and how it will help me?' The teacher was apparently stumped.

So it may come as a revelation that there are many problems which are encountered in day to day life where geometrical calculations come into play. Quite a number of them relate to playing sport. However, to the relief of many, we tend to solve these problems without consciously doing any maths at all. Regardless of how you rate your own ability to carry out complex calculations, the part of your brain that deals with co-ordination, balance and control is a computational genius.

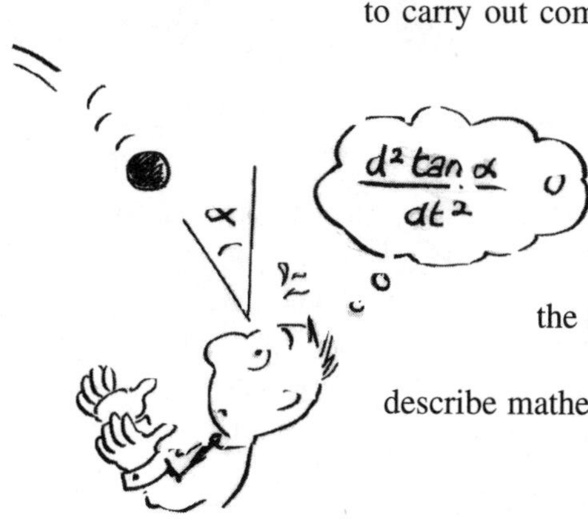

Take the process of catching a ball, for example. Did you realise that if someone throws a tennis ball through the air to you, your brain resolves a problem which would be quite difficult to describe mathematically? If you normally drop the ball, at least you now have a good excuse.

Snooker angles

Geometry crops up in many ball sports. Snooker is a classic example. Suppose a game of snooker has reached the situation shown on the left. The player is trapped in a snooker. He has to hit the white ball onto the pink, but the black ball is preventing him from a direct hit. The player needs to bounce the white ball off the cushion at the side of the table. The question is, whereabouts on the cushion should he aim to hit the white ball?

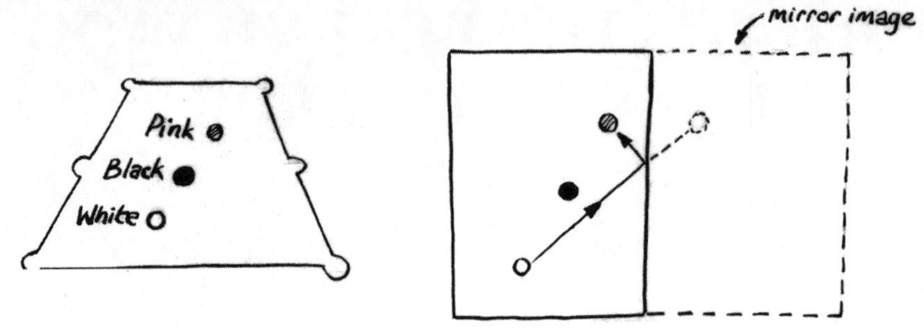

There is a simple principle which determines the point at which the white should hit the cushion. If you were to place a mirror along the top of the cushion, parallel to the side of the table, you would see a reflection of the pink ball. The player should aim the white ball directly at the pink ball in the mirror, and the cushion will do the rest.[1] (The principle behind this was understood by Heron of Alexandria around 75BC, long before the arrival of snooker.)

On the subject of snooker, you may have asked yourself as you sat watching it on TV which type of pots are easiest, and which are hardest. Have a look at the three shots facing a snooker player in the diagram below. Which of A, B and C would you be least confident of potting?

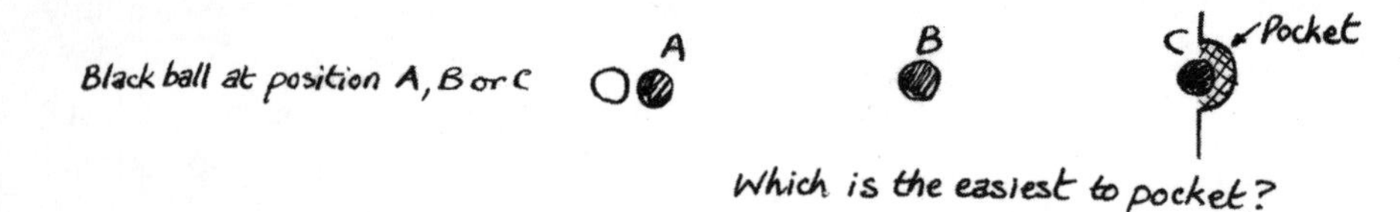

Shot B is the one that players miss most often. There is a mathematical formula that shows why this is (see opposite). The worse the snooker player is, the less true the formula becomes. As the player's accuracy deteriorates, the position at which the black is hardest to pot moves towards the pocket. In the extreme case where a player is so bad that he has difficulty using the cue at all, anything other than position A becomes almost impossible to pot. The chance of the white even hitting the black, let alone potting it, reduces to that of an outright fluke!

---

1 This ignores the effect of spin on the cushion.

Why are 'mid-way' pots the hardest?

When a good player aims the white ball in a certain direction, his alignment is extremely accurate, but it is not perfect. The error that the player makes can be expressed as an angle which can be called α. (It can be called Fred too, but why break with mathematical convention?)

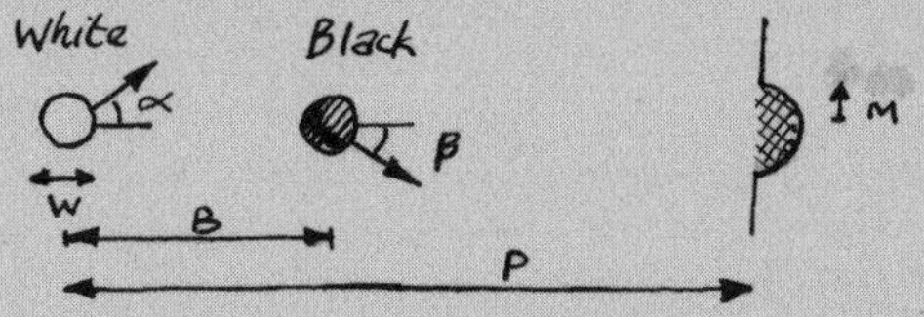

Because the white doesn't go dead straight, the black deviates at an angle we will call β. For a top class snooker player, α is tiny, and so therefore is β, and this means we can use a good approximation which says $\sin(\alpha) = \alpha$. Without going through the maths, all of this leads to a relatively simple formula for the approximate distance by which the black will miss the centre of the pocket (M):

$$M = (P-B)(B-W)/W$$

where W is the diameter of a snooker ball, P is the distance from the centre of the white ball to the pocket, and B is the distance from the centre of the white ball to the centre of the black. If B=P (the black is hovering over the pocket) or if B=W (the black is touching the white) the error is zero. The miss is at a maximum when B= 1/2 x (P + W) – a fraction over halfway from the white to the pocket.

Rugby conversions

Another sportsman who appears to successfully solve a geometrical problem is a rugby kicker.

In rugby, a team can score points by touching the ball down beyond the line for a Try. It can then add to these points by kicking a conversion. The rules for a conversion kick are that the kicker can place the ball anywhere along the line perpendicular to where the Try was scored.

Where should the kicker place the ball? If he places it on the touch line, then he is close to the posts but he cannot see a gap between them. If he places the ball in the other half of the field, he is almost front-on to the posts but they are so far away that they appear to be very close together. At kicking positions between these two, however, the angle between the posts gets bigger, and there must be a point where this angle is at a maximum. The question is, where is that point? (Strictly speaking we have to ignore things like the swerve of the ball and the distance a player is able to kick).

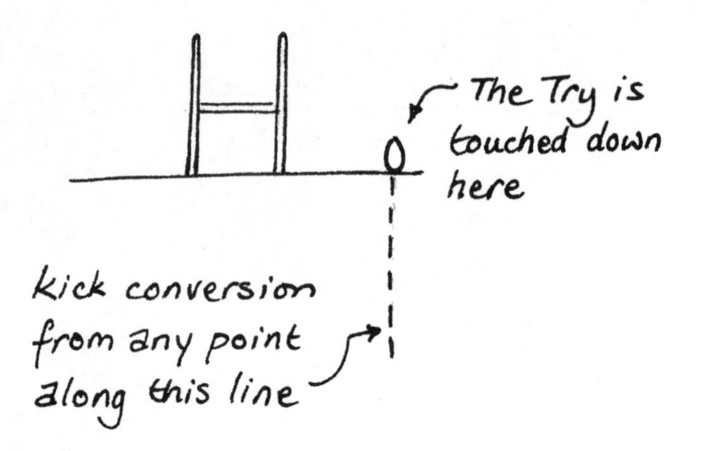

It turns out that this is a geometrical problem with a neat solution. Draw the circle that passes through the two posts and which just touches the conversion line. The point where the circle touches the line which is the *tangent*, is the best place to kick from – see the box.

There is an exception to this solution by the way. If the Try is touched down between the posts, then the kicker should place the ball as close to the posts as he is comfortable with. The angle subtended by the posts in this case gets larger the closer the ball is placed to them. The only problem now becomes lifting the ball over the bar – if the kicker placed the ball on the touchline, a conversion would be impossible!

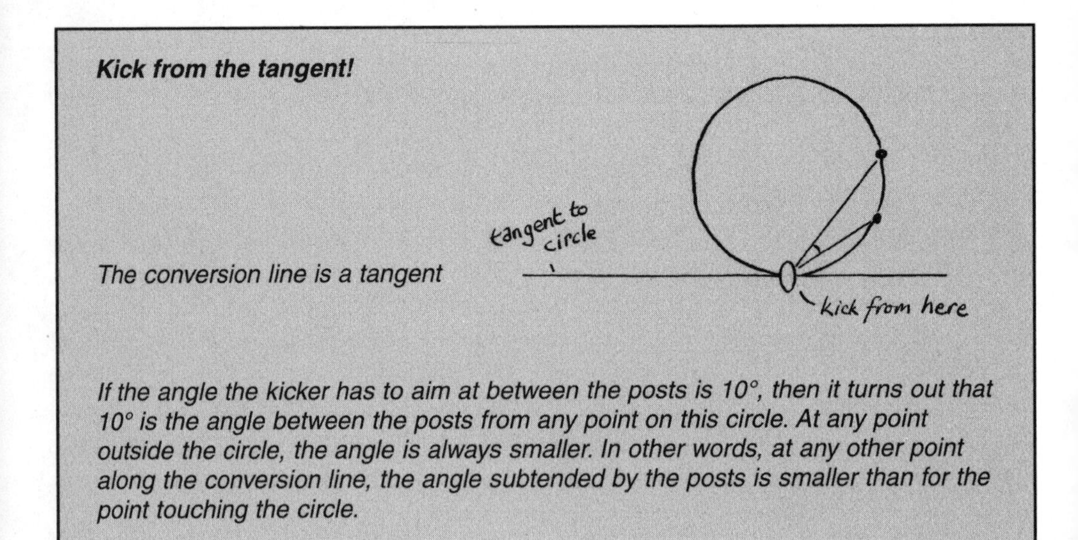

Kick from the tangent!

The conversion line is a tangent

If the angle the kicker has to aim at between the posts is 10°, then it turns out that 10° is the angle between the posts from any point on this circle. At any point outside the circle, the angle is always smaller. In other words, at any other point along the conversion line, the angle subtended by the posts is smaller than for the point touching the circle.

Looking at statues

Rugby is not the only situation where there is an optimal angle to consider. In fact the same problem arises for tourists when they try to get a good view of a raised statue.

The Statue of Liberty is a good example of this. Liberty herself is 46 metres high, but she is raised off the ground on a huge plinth which is itself 47 metres high.

If you stand at the base of the plinth and look upwards, you can see all of the statue, but she will appear stunted because the viewing angle is small. As you start to walk backward across the grass, you begin to get a better view of her because she fills a wider angle in your vision. However, this doesn't go on indefinitely. For example, from the Liberty Island ferry you get a much better side-on view of the statue, but she fills much less of your vision (you could easily squeeze the whole statue into a camera frame at that distance). So between the ferry and the base of the statue there must be an optimal point at which Liberty presents herself at the largest possible angle. It is the same as the rugby problem turned on its side.

How far back should he stand?

The optimal point is the one at which your eyes are at the tangent to the circle which also passes through the tip of Liberty's torch and her toes, as in the diagram overleaf. The formula for how far you should stand from the base of the statue assumes that the ground is flat. If this was so, then the distance would work out at roughly 64 metres, which means that the optimal point is definitely on dry land! In fact the grassy area at the foot of the statue slopes down a little, so 64 metres is not exactly right but it is close. By our estimates this puts the best position a few yards in from the edge of the grass. Alternatively, to avoid a crick in the neck, you get a pretty good view with your back to the railings next to the water.

The formula for the distance D to stand from a statue of length S and plinth height P is

$$D = \sqrt{(S \times P) + P^2}$$

The formula can be applied to other statues, too. According to our calculations Christ the Redeemer, Rio de Janeiro most fills your vision if you stand about 17 metres back from the statue in the small park area that surrounds the plinth.

Meanwhile the best place from which to view Nelson's Column in London is next to the statue of Charles I at the top of Whitehall. He is on a little traffic island, so you are safe from being flattened by a bus while you view.

Baywatch angles

A cynic might say that the millions who follow the TV series *Baywatch* are more interested in curves than in angles. There is however a classic angle problem which arises every time a life-saver sets off to rescue a swimmer in distress. It is unlikely that the drowning

swimmer is directly in front of the life-saver. He will be off at an angle. To reach the swimmer, the life-saver has to run across the sand, then dive in to the sea and swim to the rescue.

The life-saver can run much faster than he can swim, so the question is which direction should he take in order to reach the swimmer as fast as possible? On the face of it there seem to be two obvious choices:

1 Aim in a straight line for the swimmer. Everyone knows that the *shortest distance* between two points is a straight line, so this seems like a sensible strategy.
2 Aim for the point on the beach which is perpendicular to where the swimmer is stranded. This is the point which requires the *minimum amount of swimming*, which also makes sense since the life-saver runs faster than he swims.

One of these routes is always faster than the other, and which one depends on a combination of how far out the swimmer is, how fast the life-saver can swim and what angle the swimmer is stranded at. However, neither of these routes is ever the route that takes the *minimum time*. The quickest route is in fact in between the two:

This Baywatch route is exactly the same path as that of a beam of light refracting through glass. When light passes from air to glass it slows down, and the route that a beam takes from point A to point B is always the route that takes the minimum time.

There is a formula from which the correct angle (s) to run can be calculated:

$$\frac{\text{Sin (s)}}{\text{Sin (W)}} = \frac{\text{Speed of travel on SAND}}{\text{Speed of travel in WATER}}$$

So anyone caught watching *Baywatch* can now claim that they are only doing so for research purposes. 'I'm just working out whether the life-saver always chooses the optimal route for saving that scantily clad girl ...'

8

HOW DO YOU KEEP A SECRET?

Code-making and breaking isn't just for spies.

In 1587, Mary Queen of Scots was led from her room in the Tower of London to be executed at the order of Queen Elizabeth I.

Why did she meet this gruesome fate? Was it because she was a Catholic in a country which was now run by Protestants? Was it because she was part of a plot to take over the throne? Both of these factors helped to seal her fate, but what finally triggered Mary's execution was that she couldn't keep a secret.

Not that she didn't try to keep secrets. Mary used a cipher system to encrypt the messages that she sent to her supporters. Tragically the message that incriminated her was intercepted by Francis Walsingham, the head of Elizabeth's secret service, whose team had little trouble deciphering it and discovering the details they needed to stop the entire plot.

The challenge of keeping messages secret had existed for many centuries before Mary Queen of Scots. Governments and the military in particular needed systems that enabled them to pass on information that would remain secret if it did not reach the intended destination. This became an increasingly skillful science because of the ingenious methods used to decipher those messages when they fell into the wrong hands.

But although code-making and breaking is usually associated with wars, encryption has never been as important as it is today. Everyone in Western society is now involved daily in sending and receiving encrypted messages, even if they aren't aware that they are doing it. Most information transmitted electronically is encrypted, from cashpoint cards and financial transactions to e-mails and satellite television.

The electronics behind this encryption is complicated and certainly not the concern of this chapter. We are interested in the mathematical principles involved in coding and decoding, which can be traced back at least as far as the Peloponnesian Wars over two thousand years ago.

Early ciphers

The earliest known ciphers were used by the military. We know about them thanks to Herodotus and other Greek historians. One device that was used by the Spartan leader Lysander was known as a *skytale* (which rhymes with *Italy*). The person sending the message had a tapered wooden truncheon, the skytale, around which he wrapped a long leather belt. He wrote the message along the belt, then unravelled it to reveal an apparently random set of letters. He sent the belt to the recipient, who had an identical skytale and

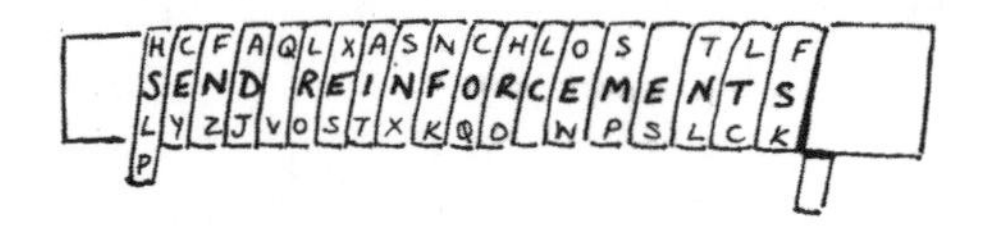

simply reversed the process to read the message. The tapering may have been the crucial part of the device. If the skytale was a regular cylinder, the coded letters in the message would be regularly spaced apart. With a tapered shape, however, the pattern of coded letters becomes irregular and therefore much more difficult to decode without knowing the appropriate dimensions of the pole and the starting point at which the belt is placed.

Julius Caesar was fond of a different method of ciphering. He used a simple substitution system in which he shifted each letter of the alphabet along by three places. The Roman alphabet used 23 letters. They had no J, U or W so his key would have been:

Plain Text	A	B	C	D	E	F	G	H	I	K	L	M	N	O	P	Q	R	S	T	V	X	Y	Z
Cipher	D	E	F	G	H	I	K	L	M	N	O	P	Q	R	S	T	V	X	Y	Z	A	B	C

Presumably he signed his letters 'FDHXDV'. This method of ciphering has been known as the Caesar system ever since.

Codes and ciphers

Secret messages belong to two categories – codes and ciphers. Codes used to be popular with diplomats. These involved translating whole words or phrases into other words or even symbols. For example 'The King' might always be represented as 'My Aunt' in code. The recipient of a message carried a code book for interpreting the message.
Ciphers were more popular with the military. Ciphers involve the translation of individual letters into other letters or symbols, a substitution cipher. In sophisticated ciphers, the order of the letters is also jumbled up. This is known as a transposition cipher.
Modern cryptography almost exclusively involves ciphering.

Emperor Augustus clearly liked this idea, because he adopted a similar system, although a shift of three was too complicated for him. His method used only a single displacement of the letters. AVGVSTVS became BXHXTVXT. Since Julius Caesar's technique had already been widely publicised (even by Caesar himself) it is surprising that Augustus' messages had any security at all.

There is a mathematical way of describing a Caesar cipher. Take the following example using the modern alphabet:

Plain	A	B	C	D	E	F	G	H	I	J	K	L	M	N	O	P	Q	R	S	T	U	V	W	X	Y	Z
Cipher	D	E	F	G	H	I	J	K	L	M	N	O	P	Q	R	S	T	U	V	W	X	Y	Z	A	B	C

To begin with the letters of the alphabet are translated into the numbers 1 to 26.

Letter	A	B	C	D	E	F	G	H	I	J	K	L	M	N	O	P	Q	R	S	T	U	V	W	X	Y	Z
Number	1	2	3	4	5	6	7	8	9	10	11	12	13	14	15	16	17	18	19	20	21	22	23	24	25	26

The same is done for the cipher:

Letter	D	E	F	G	H	I	J	K	L	M	N	O	P	Q	R	S	T	U	V	W	X	Y	Z	A	B	C
Number	4	5	6	7	8	9	10	11	12	13	14	15	16	17	18	19	20	21	22	23	24	25	26	1	2	3

The sequence fits the equation **Cipher = Plain + 3**, until the 24th letter, X. Letter X is translated not into 27 but into 1 (which is A). This is easily put right by using modulo arithmetic, sometimes known to children as clock arithmetic. In this case, every time a number gets above 26, multiples of 26 are removed until the answer is between 1 and 26.

So the formula for this cipher is **Cipher = Plain + 3 (mod 26).**

Adding 23 to the cipher restores the original message, so it is possible to have a formula for deciphering the code as well. This is the 'key' with which the cipher is unlocked: **Plain = Cipher + 23 (mod 26).**

These examples are so simple that it is hardly necessary to turn the cipher into a formula. That changes when ciphers become more complicated, as you will see. In fact modulo arithmetic has a vital part to play in the more complex ciphering systems that are used today. But first let's take a quick look at some more sophisticated methods of ciphering.

Clock (or modulo) arithmetic

Children learn clock or modulo arithmetic when they learn to tell the time. What comes 13 hours after 4 o'clock? Not 17 o'clock but 5 o'clock. Multiples of 12 are taken away from 17 until the number is between 1 and 12.
This is the basis of the puzzle about the man who is exhausted and so goes to bed at 9 o'clock and sets his wind-up alarm for 10 o'clock the next morning. How much sleep does he get? Children usually answer 13 hours, or thereabouts, whereas you of course immediately spotted that he only gets one hour's sleep before the alarm goes off at ten o'clock that evening.

Substitution ciphers

Mary Queen of Scots did at least have a more sophisticated cipher than Caesar. Her method involved shuffling the alphabet around, and also inserting dummy letters and symbols to confuse the issue. The letter B might represent a C, while E would be represented by a Z. But this still made the cipher vulnerable to frequency analysis (see the box).

One technique that begins to overcome the frequency problem is to regularly shift the cipher code that is being used. The following sentence has been encrypted using just such a technique and you might like to have a go at deciphering it:

> This sentencf ibt cggp gpetarvgf da wukpi vjg ngvvgt e vq ujkhv vjg coskdehw eb rqh hdfk wlph lw lv klw

The translation of the sentence is in the footnote opposite.[1]

The words in this encrypted sentence are still recognisable by their length, however, which is why encrypters like to ignore blanks and to group letters in fives. Hence the above becomes the much less readable:

thiss enten cfibt cggpg petar vgfda ... and so on.

What is convenient about this type of cipher is that the instruction for deciphering it is so simple. All that the recipient would need to know would be the instruction 'C1', meaning C is the shift letter and 1 is the amount by which the alphabet shifts each time. The ideal code is one that is simple to crack with the key, but impossible without it. Often in past wars spies were uncovered because they were carrying complicated decryption systems stored in notebooks – and notebooks are much easier for the enemy to find than a simple instruction like C1.

Frequency analysis – the Scrabble factor

Walsingham used what is known as 'frequency analysis' to crack Mary's code. Some letters appear far more often than others in English documents. E is by far the most frequent letter followed by T. In fact in a long document, the pattern becomes quite predictable:

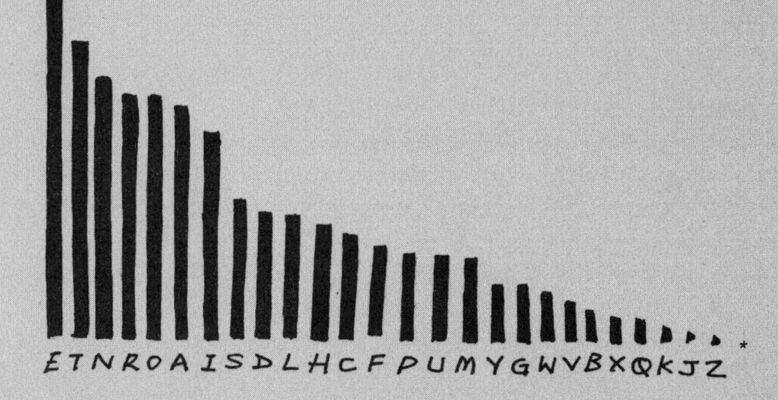

These were the frequencies of letters in a large sample of government telegrams. This uneven distribution allows a simple substitution code to be broken quite quickly, since the letters E, T, N, R, O and A represent over 50 per cent of all letters used. This is why subtler techniques are required in which an E is not always represented by the same symbol. The Germans' ENIGMA machine in World War II shifted the cipher after every letter, resulting in a frequency distribution which was almost a horizontal line. This is what made it so hard to crack. Nevertheless, frequency analysis and the search for any sort of unusual pattern remains one of the main tools of the codebreaker.

** Notice the link between this letter distribution and the points in the game of Scrabble. The less frequent the letter, the higher its Scrabble points, more or less. The main anomaly appears to be U, which only scores 1 point. No wonder those Us are so hard to get rid of ...*

1 This sentence has been encrypted by using the letter c to shift the alphabet by one each time it is hit.

The matrix substitution

The cipher becomes even harder to break if it is not single letters but double letters that are being encoded. In a simple substitution cipher, the message 'A CAB' might be translated into D FDE. But more subtle is to use a grid which enciphers pairs of letters. A CAB becomes AC AB and is then encrypted using this:

Second letter in pair	First letter in pair: A	B	C	D	E . . .
A	DI	EK	FM	GO	etc
B	GP	HR	IT	..	
C	JW	KY	LA	..	
D	..	..	..		
⋮					

AC becomes JW. AB becomes GP. The result is that A CAB has become J WGP, and the letter A has appeared as two different enciphered letters, J and G. Instead of 26 symbols to unscramble, there are now 26 x 26 or 676 pairs of symbols. This makes life tough for the codebreaker, but unfortunately it does mean that the authorised decoder has a big matrix to refer to. Wouldn't it be better if he could have something simpler as his key? How about a tiny matrix that does the same thing? There is one!

By turning the letters A to Z into the numbers 1 to 26, the code grid above can be reproduced using two formulae. In these formulae P_1 and P_2 represent the pair of original plain text letters as numbers (e.g. A, B is 1, 2) and C_1 and C_2 are the enciphered letters:

$$C_1 = 1 \times P_1 + 3 \times P_2 \pmod{26}$$

$$C_2 = 2 \times P_1 + 7 \times P_2 \pmod{26}$$

Try it for the pair of letters AB which in number form are 1, 2.
The first letter becomes 7 (letter G), and the second becomes 16 (letter P). So AB becomes GP when it is enciphered.

The reverse process is carried out using the appropriate decoding formulae, which in this case are:

$$P1 = 7 \times C1 + 23 \times C2 \pmod{26}$$

$$P2 = 24 \times C1 + 1 \times C2 \pmod{26}$$

Again we can test this out. G is 7 and P is 16. This gives:

$$
\begin{aligned}
P1 &= 7 \times 7 + 23 \times 16 \pmod{26} \\
&= 417 \pmod{26} \\
&= 1 \\
&= \text{'A'}
\end{aligned}
$$

$$
\begin{aligned}
P2 &= 24 \times 7 + 1 \times 16 \pmod{26} \\
&= 184 \pmod{26} \\
&= 2 \\
&= \text{'B'}
\end{aligned}
$$

Hey, it works! The decoder can now carry around the key in the form of the four numbers in the formulae, 7,23,24,1 (which he can write in the form of a matrix if he wants – see the box overleaf). Mind you, the calculations would be fairly horrendous, which is where computers come in.

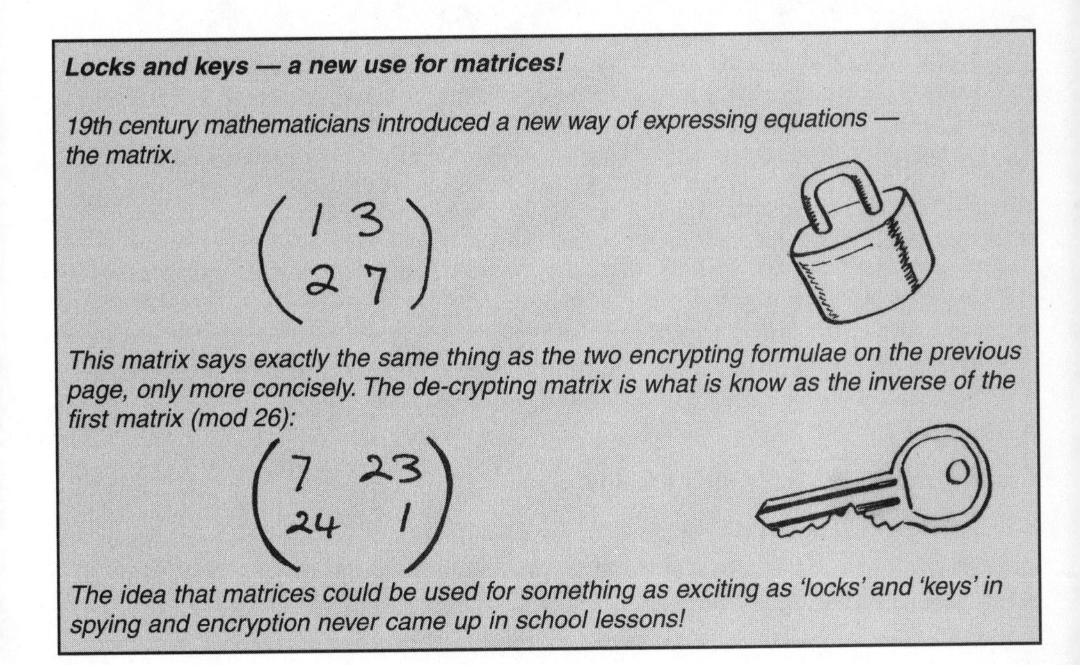

Locks and keys — a new use for matrices!

19th century mathematicians introduced a new way of expressing equations — the matrix.

$$\begin{pmatrix} 1 & 3 \\ 2 & 7 \end{pmatrix}$$

This matrix says exactly the same thing as the two encrypting formulae on the previous page, only more concisely. The de-crypting matrix is what is know as the inverse of the first matrix (mod 26):

$$\begin{pmatrix} 7 & 23 \\ 24 & 1 \end{pmatrix}$$

The idea that matrices could be used for something as exciting as 'locks' and 'keys' in spying and encryption never came up in school lessons!

Transposition ciphers

The trouble with all of the previous examples of ciphers is that the order of letters in the message has not changed. It can be a lot more baffling if the order of the letters is shuffled around like an anagram. A very simple transposition technique involves writing out a message on a rectangle. For example, WE HAVE RUN OUT OF BEER is written as:

W E H A V E
R U N O U T
O F B E E R

It is then enciphered by reading down the columns as:

W R O E U F H N B A O E V U E E T R

The dimensions of the rectangle determine the shuffling order, and these dimensions can be transmitted at the start of the message. For example DEAR (four letters) MOTHER (six letters) could be used to indicate a 4 x 6 rectangle. Variants of this technique were used by the North in the American Civil War.

> ***How the North beat the South at ciphers***
>
> *Cryptography played a big part in the American Civil War, and it has to be said that the Union in the North were a lot smarter than the Confederates in the South. The Union under Abraham Lincoln used transposition ciphers and made sure they changed their keys regularly. The Confederates found these impossible to crack – in desperation they even published intercepted Union messages in their newspapers, asking if the public could help them to find a solution. Meanwhile the Confederates' own codes were less well co-ordinated, and some generals even resorted to Julius Caesar's method. Needless to say, many of these messages were deciphered by the enemy.*

When transposition is combined with substitution, breaking the code starts to become a real headache. But not as much of a headache as the trapdoor ...

Trapdoors and the truly uncrackable

You have now met most of the main principles behind cryptography. To understand how modern cryptography works, you need to imagine everything that has just been said being made a billion times more complex. That's what computers have done for us.

In recent years, cryptographers have devised what some regard as the almost perfect cipher – easy to create and decode if you have a computer, but impossible to crack even with the most powerful machines on the planet.

For years, mathematics professors worked away in ivory towers on the subject of *number theory*. This area of mathematics belonged to the world of pure academia. There were almost no known practical applications for this abstract subject. Not until 1976, that is. In that year number theorists Diffie, Hellman and Merkle announced to the world that they had discovered what they called a trapdoor function which was perfectly suited to cryptology.

Trapdoor ciphers are so called because of their 'one way' nature – anyone can easily fall through a trapdoor, but getting out is far harder, unless you have the right ladder.

Here is a mathematical trapdoor. Measure how long it takes you to do the following two problems:

Problem 1 What is 13 x 23 ?
Problem 2 What two numbers multiplied together give 323?

With a calculator, problem 1 can be solved instantly to get 299. Problem 2 takes a lot longer to solve, because it requires trial and error. The answer is 17 x 19. It is a unique answer because 17 and 19 are prime numbers, i.e. they are divisible by no number except 1 and themselves.

Multiplying together 17 and 19 is far easier than working out the factors of 323. So imagine if the two prime numbers chosen were 100 digits long instead. Multiplying them together would take a computer just a few seconds. But working out what those two prime numbers are from knowing their product could take the computer millions of years because of the vast number of trials it would have to make. This is the secret of the trapdoor.

To encode using a trapdoor cipher, the cryptographer first secretly chooses two huge prime numbers, each with (say) 100 digits. These two numbers are multiplied together to create an even huger number about 200 digits long which we will call M. Finally the cryptographer finds a third prime number. It needn't be big – 101 would do. We will call this number P.

The original message is translated into a single number. For example if A = 01, B = 02 and so on, then it turns out that the message 'SEND MORE MONEY' becomes '1905140413151805131515140525'. This now has to be encrypted – which is where things start to become complicated.

The message number is raised to the power of P, but using modulo arithmetic. Earlier examples had encryptions working in modulo 26. That was nothing compared with the trapdoor function, which works in modulo M. (M is the 200 digit number that was calculated using the big prime numbers). The result will therefore be a number with around 200 digits. The 200 digit result represents the original message, but in a form which is completely unreadable (remember, the original message SEND MORE MONEY only had 26 digits, so it looked nothing like the enciphered message).

At this point, the sender says to the code-breaker 'OK, crack this one pal!' To reverse the calculation, the code-breaker needs to find the two prime factors of M. Even with the world's most powerful computer it could be a million years before he does so. Yet if the original two primes are known, the decoder can extract the message extremely quickly with a computer. Those primes represent the ladder which allows a return through the trapdoor.

Trapdoor functions are perhaps the elite of a number of mathematical techniques that are used to ensure that nobody can illegally read an e-mail, a bank balance or a satellite TV broadcast. They have turned what was once the realm of the gifted amateur into a subject to be grappled with by only the very best of mathematical brains.

They have rather taken away the fun, though.

9

WHY DO BUSES COME IN THREES?

Travelling without a car leads to all sorts of conundrums.

Everybody knows that if you want to catch a bus, you spend ages waiting and then three will come at once. That at least is the urban myth which is popular enough to become the title of a book. According to mathematicians, however, it really is a myth. Buses don't usually come in threes, they come in twos, and the reason why this is so can be found in the box over the page.

For the time being, however, let's suppose that buses *do* come in threes. If it is true, then a well known commuter's nightmare may not be a nightmare after all.

Maybe you are one of those unlucky people who always misses a bus when you have an important appointment to get to. You might imagine that it could never be a good thing to have just missed the bus. However, if buses come in threes, then just missing a bus can mean that you can expect to get to your destination *faster.*

How can this be? How might it actually be a good thing if you missed the bus?

In order to investigate the bus phenomenon, we'll have to devise what is known as a mathematical model – a simplification of the real situation with some invented figures. As long as the assumptions are sensible, models can provide an insight into the way things work.

Missing the bus can be good

Suppose that buses leave the depot every fifteen minutes, but by the time they reach your stop, the buses are bunched in threes. For the sake of argument, let's also say that the three buses in a bunch are only a minute apart.

Since three buses leave the terminus in any spell of 45 minutes, the diagram shows that the gap between bunches of three buses must be 43 minutes.

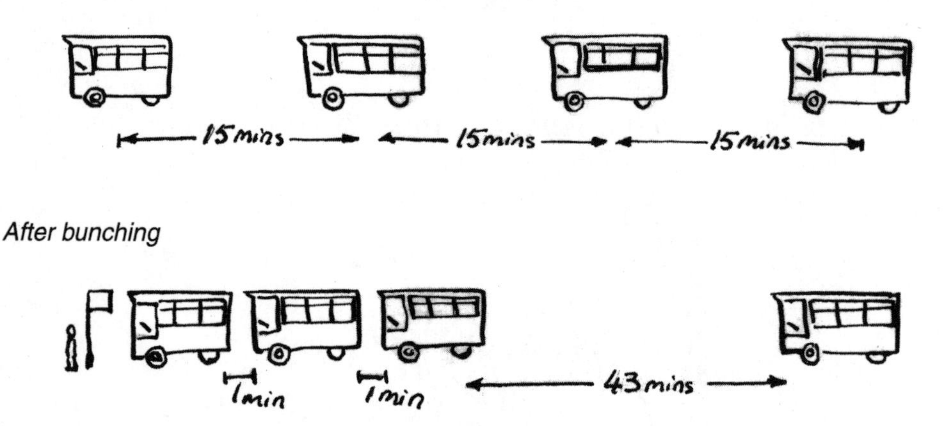

Let's now suppose you have just seen a bus leaving your stop. You don't know which of the three bunched buses it is. It is equally likely to have been the first, the middle or the last bus. If it was the first or second bus you only have to wait a minute for the next one. However, if it was the third bus, you have to wait 43 minutes.
This means that the average time that you have to wait for the next bus is:

$$\frac{1 \text{ minute} + 1 \text{ minute} + 43 \text{ minutes}}{3} = 15 \text{ minutes}$$

But what if there is no bus at the bus stop when you arrive? In other words, what happens when you *haven't* just missed the bus? This means you have arrived in one of the gaps between buses. You may have caught a one minute gap, but the chances are 43/45 that you have found the long gap. And you could have arrived anywhere in the long gap – at the very start with 43 minutes to wait or at the end with the next bus about to arrive. On average, therefore, the time you have to wait is now (43+0)/2 = 21.5 minutes. Even adjusting for the small chance that you arrived in a one minute gap, if you don't see a bus leaving from your bus stop you actually spend an average of over five minutes longer waiting for your bus than if you do see one leaving.

That is why just missing a bus can shorten your overall journey time.

This curious result does rely on buses bunching in threes, however. And as the box opposite shows, buses are much more likely to bunch in twos than in threes. If buses bunch in twos, it turns out that just missing a bus has no effect on your waiting time.

The worst situation for the passenger who misses the bus is if the buses don't bunch at all. In this case, missing the bus definitely means waiting 15 minutes, while not seeing a bus now means an average wait of only 7.5 minutes. But then at least if you see a bus leaving you know that they are running today ...

Do buses REALLY come in threes?

The reason why buses bunch is nothing to do with incompetent planning by the bus companies. Bunching is a simple fact of life. Even if buses leave the depot every fifteen minutes on the dot, the passengers arriving at bus stops are not so consistent. They arrive much more at random. It is likely that at some point on a bus route there will be a sudden burst of passengers arriving, and of course they all have to pay to get on. This behaviour slows the bus down, and so allows more passengers to collect at the next stop.

Meanwhile the bus behind is now not only closer to the one in front of it, but also, because there is now less time for passengers to arrive between buses, there are fewer passengers for the second bus to collect. So the second bus progresses even faster. The buses have now entered a vicious circle which will almost certainly lead to the second one catching up with the first, whereupon the two will complete their journey in partnership. Hence the tendency for buses to bunch in twos.

The further along a route that a bus goes, the more likely it is to bunch with another bus. If bunches of three happen at all, they are most likely to appear towards the end of a long bus route. They are also more common when buses start their journeys close together, in other words on bus routes with a frequent service. How ironic that the 'best' bus routes can be the ones which attract the worst reputation for bunching.

Buses and trains in two directions

Related to the problem of bunching buses is another curious phenomenon that could really happen. Suppose your bus stop is very close to the end of the bus route where the buses turn around and set off back the way they came. You notice that when you turn up at random to catch your bus, you nearly always see one of your buses going in the opposite direction before you see one coming in your direction. It feels like a conspiracy. Should you write in and complain? The explanation for the uneven bus problem is similar to the one for the flower story in the box below.

Let's invent some times for your bus route. Your stop is only a minute away from the end of the route, and the entire circuit of the bus route takes fifteen minutes. This means your bus comes every fifteen minutes. When you arrive at your stop, you are either in the thirteen minute gap while the bus is on the long part of the route, or the two minute gap while the bus goes to the end of the route and back.

As long as your arrival is random, this means there is a 13 to 2 chance that you are in the long gap, and that the first bus you see will be your bus on the other side of the road before it reaches the terminus. In fact, it doesn't matter how many buses are on your route,

Why did Sarah get all the flowers?

Phil has two girlfriends, both of whom he visits by train. Becky lives north of the town and Sarah lives south. Since he can't decide which of the two he should visit he decides to let chance decide. He arrives at a random time at the station every day, and if the north train arrives first he visits Becky, while if the southbound one arrives he visits Sarah. After a month Phil starts to feel fate is trying to tell him something, since he has only visited Becky twice, while Sarah has had 28 visits. What is the explanation?
The answer has nothing to do with the frequency of trains. They run just as frequently north as south. The explanation is in fact very simple. The trains heading south arrive at Phil's station on the hour, and at 15, 30 and 45 minutes past. The trains heading north arrive at 01, 16, 31 and 46 minutes past the hour. So if he turns up at random, he is much more likely to be in the long gap before

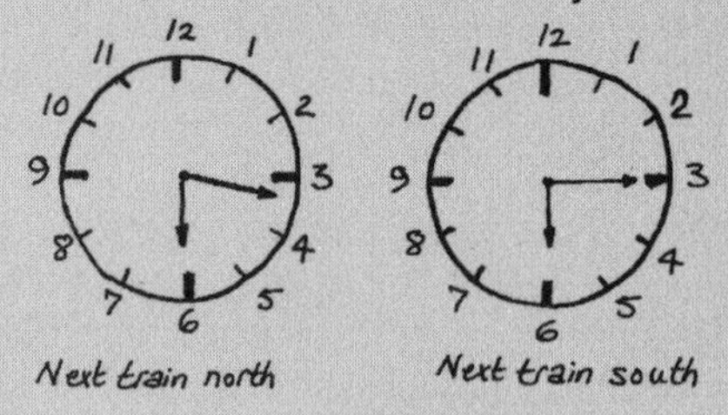

the south train than for the short gap for the north train. (Fourteen times as likely, in fact.)

this ratio still applies if buses come every 15 minutes. So it might appear to you that buses are serving the opposite side of the street more often than they serve yours, when this is not actually the case.

How fast should you run in the rain?

There has been a lot about waiting for buses and trains in this chapter. There are of course those occasions when public transport fails to turn up altogether and you have to walk. The problems pile up when it then starts raining and you don't have your umbrella with you.

The age old question is whether you should run or walk. If you run, think of all those extra raindrops that you are colliding with which you would otherwise have missed. On the other hand, if you walk you will be out in the rain for longer, catching lots of rain on your shoulders. Some serious mathematical thought has gone into this question over the years. The conclusion has been that, to stay as dry as possible, you should run as fast as you can. Common sense probably told you the same thing.

However, there is a surprising twist to this problem. The standard solution assumes that the rain is falling vertically. What happens if there is wind and the rain is falling at an angle?

When rain falls vertically and you are standing still, the drops only hit your head and shoulders. If there is wind coming from behind you, however, then some of the rain will also hit your back *even when you are standing still*. It's as if the rain is falling horizontally as well as vertically. The rain has a *horizontal speed*. The surprising twist is that when the rain is coming from behind you, it is sometimes better to walk than run. However, this only applies if you are able to move faster than the horizontal speed of the rain.

Going fast can sometimes be a disadvantage

Why it is sometimes drier to walk in the rain than run

Suppose that rain is falling at an angle K and is coming from behind the pedestrian.

To simplify the calculation, let's assume that the pedestrian is a rectangular block of wood. (Strap his legs and arms together, lop his head off and the resemblance is striking.) There are seven factors that we have taken into account:

V is the speed at which the rain is falling

K is the angle at which the rain is falling

D is the density of the rain (in kg per cubic metre)

A_t is the area of the pedestrian's top

A_f is the area of the pedestrian's front

H is the distance to the pedestrian's destination

V_p is the speed at which the pedestrian runs

There isn't space to put in all the algebra, but the way we produced the formula below was to work out separately how much rain hits the front and the top of the pedestrian and then added these two together.
The total rain in kilograms falling on the pedestrian as long as he is travelling at least as fast as the horizontal speed of the rain is:

$$DA_fH + \frac{DHA_tV\cos K}{V_p}\left(1 - \frac{A_f}{A_t}\tan K\right)$$

The important part of the formula is the bit in the brackets. If $A_f/A_t \tan K$ comes to more than 1.0 then the right hand part of the formula becomes negative, i.e. the rain hitting the pedestrian is reduced. The pedestrian can control his speed V_p, so to stay dry as possible, the pedestrian should go no faster than the horizontal speed of the rain if $A_f/A_t \tan K$ is greater than 1. (Take a deep breath!)
The ratio of a person's front area to their top is usually about 5.0. Since the tangent of 15° is about 0.2, this means that if rain is falling at an angle of more than 15°, the driest way to get home without an umbrella is to move at the same speed as the horizontal speed of the rain.

The formula for how wet you get is given in the box opposite. However, to sum up, the conclusion is this:

If you are a person of typical build and the rain is coming from behind you at the speed of a gentle walking pace, you will be hit by less rain if you amble along than if you run full pelt.

In other words, there are circumstances when it is better to walk than run!

The reason for this apparently odd outcome is that under the conditions described, if you run faster, the additional rain hitting your front is greater than the loss of rain hitting your head by taking less time to make it home!

Of course by the time you've worked all this out you will be wetter than if you had not known about the formula at all!

10

WHAT'S THE BEST WAY TO CUT A CAKE?

Why four o'clock can be the time for some mathematical headaches.

Afternoon tea is a popular ritual in England that is so routine, it might seem hard to believe that it can have anything to do with math. Yet buried within the cosy four o'clock ritual are some intriguing problems which mathematics can help to resolve.

Let's start with the central activity, pouring the tea. One of the big problems with tea is getting the temperature right. Either it is scaldingly hot when you first put the cup to your lips, or else it has become horribly lukewarm by the time you get to the final sip. One mathematical question of special interest to students of physics, as well as to the people running the village fête, is how best to retain the warmth in a cup of tea. Should you put in the milk first and then the tea, or should you pour in the tea and then the milk? It does make a difference, although in which direction is often a matter for debate.

What effect putting the milk in first has on the taste and your social status is another question entirely.

Milk or tea first?

The generally accepted view is that putting the milk in first will retain the heat in a cup of tea for longer. The reason for this is that the rate at which an object loses heat depends on the difference between its own temperature and that of its surroundings. Hot tea without milk starts at a higher temperature and therefore loses its heat faster, although the difference is so small that it is hard to detect.

Putting the hot tea in first can also be a problem for cheap crockery. A sudden change of temperature can cause thick china to crack. Hot tea is less likely to crack a delicate, thin cup because the heat spreads rapidly to the cup's outer surface. This is one reason why putting tea in first became popular with the wealthy classes of society. It made the statement 'we have the kind of crockery that doesn't crack with hot tea.'

Minimising the number of cuts

So much for the tea. What about the cake? The simple act of cutting a cake has within it a whole myriad of mathematical principles, many of which have been set as puzzles since Victorian times.

Take the following, for example. You have a birthday cake to divide between eight children. How can you cut it into eight equal pieces using only three straight cuts and without moving any pieces of the cake?

The answer requires a little lateral thinking (and lateral cutting). Make two cuts from above and a third one horizontally through the middle of the cake.

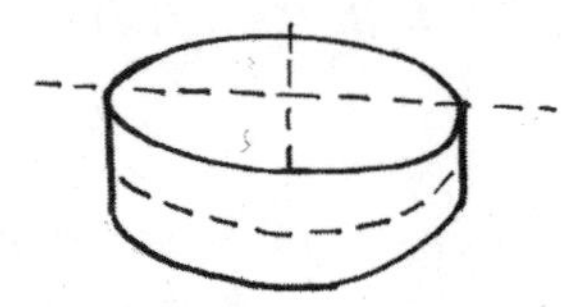

This is all very well for a puzzle, but if the real cake has icing and marzipan, the recipients of pieces from the lower half may be very upset. In fact a different problem arises if you have a cake which is iced all over. Suppose this cake is square and you need to cut it into a number of identical pieces, each with the same amount of icing? Two, four or eight pieces can be cut by simply halving the cake and its pieces. But how about an odd number of pieces? There is a slightly eccentric method that works regardless of how many people you are cutting for. All you need to do is mark out the perimeter into equal lengths. If your number of guests is seven, for example, mark out the perimeter of the square into seven equal lengths.

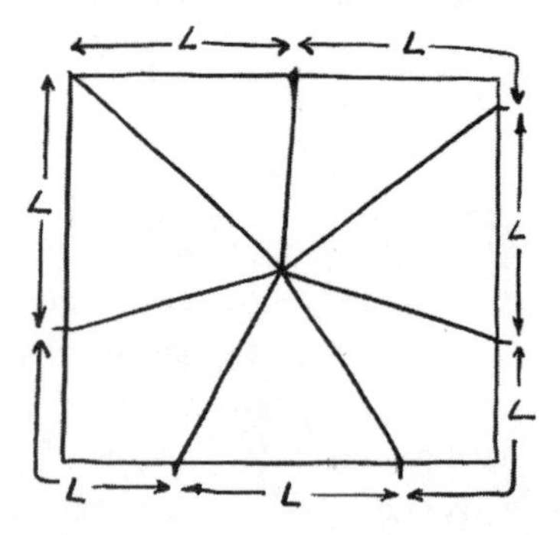

Now locate the centre of the cake and make cuts from the perimeter marks to the middle as shown. These seven pieces are of equal volume and have the same amount of icing, as explained in the box below.

This method works for any cake that is the shape of a regular polygon. If you want to cut a triangular cake into ten pieces, you mark out the perimeter into ten equal lengths. Store this knowledge up, because one day somebody will produce a triangular cake ...

Fair play

Where children are concerned, dividing a cake equally can be quite an issue. They are only too quick to complain when they think the cake has been unfairly divided. Adults will rarely complain out loud, although they too are probably seething underneath. So how do you ensure that a cake is divided fairly? Let's assume that it is a sponge cream bun, and that there is therefore a strong incentive to completely divide out the cake first time, with no seconds.

Take a simple problem to start with. Tom and Katy have a cream bun and their mother wants them to have half each. Neither of them believes that their mother is capable of fairly dividing it in two and they both think they will lose out. How can their mother ensure that both children *believe* they have been treated completely fairly?

Half base times height – it's a piece of cake (if you like algebra)

The proof for the 'perimeter method' of cutting a cake is based on the simple property of a triangle. Suppose the cake is 10 inches square and you want to divide it into five equal pieces, each with the same amount of icing. The cake below has been cut into five segments after marking out the perimeter. Each piece has been laid out on the right. The pieces that include a corner of the square cake have been divided into two triangles, and the perimeter points a, b, c, d, e, f, g and h labelled on the left have been marked on the right as well.

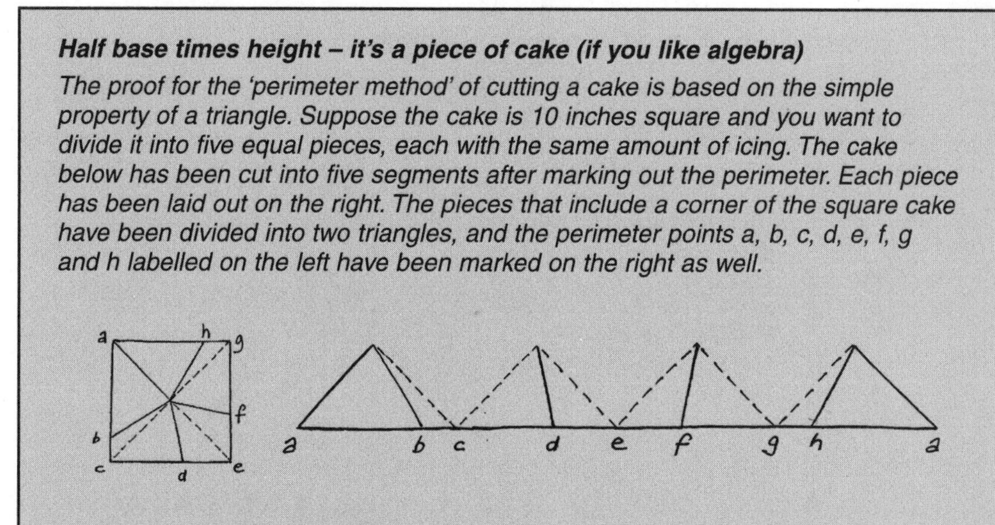

To work out the sizes of the slices, you only need to know that the area of a triangle is given by the formula Area = ½ base x height.
Each triangle has the same height, 5 inches (half the length of the side of the original square cake). Each slice of cake has one fifth of the perimeter, which is 8 inches. So the area of each whole slice is $1/2$ x 8 x 5 = 20 square inches.
The perimeter method works regardless of the number of slices you divide the cake into.

Fair shares

The answer is to give the knife to Tom, ask him to divide the bun and then ask Katy to choose which piece she wants. Tom will divide the bun so that he believes the halves are identical, while Katy will choose the piece which she thinks is bigger. Incidentally, this has an interesting consequence. Tom thinks he has been left exactly half and Katy has taken what she thinks is slightly more than half. Tom's 'half' plus Katy's 'more than half' adds up to more than one. If the logic of this maths is taken to its conclusion then, in the view of the children, there is more bun at the end than there was at the beginning! This is good news for any parent who wants to keep children content.

HAM SANDWICH THEOREM

The Ham Sandwich Theorem was developed by two mathematicians, John Tukey and Arthur Stone. It states that the volumes of any three solids can be bisected simultaneously by a single plane no matter where these three solids are placed or no matter what size or shape they are. This theorem can be applied to any sandwich. The three solids become two slices of bread and the filling, whilst the bisecting plane becomes the knife cut.

What this means is that no matter what shape the slices of bread are (and they can be different from each other) or what shape the filling is, a single knife cut can divide the sandwich into exactly two halves. Unfortunately the theorem does not give the precise position of the knife cut – only that such a cut exists!

The problem becomes more complicated if there are three children. Let's suppose Emma has now arrived. The simple strategy is to ask Tom to divide the bun into three, then let Katy choose first and Emma second. Unfortunately, while both Katy and Emma will think they have done better than Tom, Emma may feel that Katy was able to take the biggest piece.

This introduces the mathematics of envy. The problem has been investigated in some detail by a number of mathematicians, who have found ways of dividing a cake three ways so that each recipient believes they have the best piece. Messrs Brams and Taylor tackled the problem for four people as well. They produced a remarkable 20 stage approach which guarantees that a cake can be divided such that each person believes that they have chosen the biggest piece. The big disadvantage of the method is that it involves shaving pieces off a cut slice. Few people would have the patience to follow it through, and with a gooey cake it would create a terrible mess.

However, Brams and Taylor found that their procedure could be applied to things other than cake. These include dividing territory after a war, allocating the possessions between a divorcing couple or even dividing up an inheritance. All of this proves that cakes and sandwiches are a good starting point for investigating the mathematics of justice. However, they are also a good introduction to the study of guilt.

Guilt and biscuits

Suppose that you and four neighbours are invited to tea by Mrs O'Callaghan at number 27. When you arrive, she brings in a pot of tea, and a plate on which there are five biscuits, four of them chocolate and one of them plain. You suspect that most of the neighbours are chocolate biscuit lovers.

The plate of biscuits is placed on the table and as you are all chatting, the first three neighbours help themselves to a chocolate biscuit each.

You look at the plate, with one chocolate and one plain biscuit, and think to yourself 'If I take the plain biscuit, I won't enjoy it, but then I won't feel guilty. On the other hand, if I take the chocolate biscuit I will enjoy it but I *will* feel guilty – what shall I do?'.

The question is, should you really feel that guilty for taking the last chocolate biscuit? After all, if the first person had taken the plain biscuit then the next four would only have had chocolate biscuits to eat. So maybe the first person should take some of the guilt for leaving you with this predicament. Similarly for the second and third person.

Within this problem is the mathematics of guilt, which has some connection with probability. If 80 per cent of people prefer chocolate biscuits to plain, then when the first neighbour helps himself to a chocolate biscuit, the chance that every other neighbour will want a chocolate biscuit is only about 40 per cent (this is 0.8 x 0.8 x 0.8 x 0.8, similar to the birthday calculation in Chapter 6). So the first neighbour needn't feel too guilty.

By the time it gets to you, however, with one chocolate biscuit and one plain on the plate, the chance that the other neighbour wants the chocolate one has risen to 80 per cent. No wonder you feel guilty. But then, each of the other neighbours has progressively added to the crime, and so none of the chocolate biscuit-takers are entirely innocent.

Several tactics have been suggested for overcoming the chocolate biscuit guilt problem. The first is to pick up the plate and offer it to everyone to start with, asking if anyone wants a plain biscuit. If anyone takes the plain biscuit, then you are free to take the chocolate biscuit you desire without any guilt. The disadvantage of this approach is that nobody likes to take the last one of anything, chocolate or otherwise. The plain biscuit has suddenly become the source of guilt itself.

An alternative approach is for you to announce that you aren't hungry, so everyone else can share the biscuits between them. Some would argue that this gesture of selflessness is one that will command the respect of everyone else in the room, and can only enhance goodwill between the people of this planet. Others would regard it as a wet cop out.

That leaves one final strategy, which is to lump all of the guilt on Mrs O'Callaghan. 'Excuse me, but there are five of us and only four chocolate biscuits'. This is likely to solve the problem in the short term as Mrs O will nip around to the corner shop, but it might be the last time you are invited for tea.

Every potato chip may have a twin

Everybody knows that a potato chip starts its life as a thin slice of potato. But is every chip a different shape? If you pick up a couple of ordinary potatoes what would you think were the chances of being able to find two identical chips, one in each potato?

The surprising answer is that you can always find two potato chips with identical shapes in the two potatoes, even though no two potatoes are the same.

To see why, imagine the potatoes were not solid and could be intersected like soap bubbles.

The circumference of the left hand potato at the intersection is precisely the same shape as the circumference of the right hand potato. This means they must have a chip in common at this intersection. In fact there are infinite such intersections, and so an infinite number of shared chips in every two potatoes! (This is strictly true only if the potato chips are infinitely thin.)

11

HOW CAN I WIN WITHOUT CHEATING?

Almost everything in life can be analysed as a game.

In 1947, Stephen Potter introduced the world to the delights of gamesmanship, which he described as 'the art of winning at games without actually cheating'. The secret of gamesmanship was to undermine your opponent by playing psychological games with him. One example he gives is of a gamesman who is playing golf and losing to his opponent. As the opponent walks up to play his next fairway shot, our hero uses a well-known ploy:

Gamesman: ... Do you mind if I come round to this side of you? I want to see you play that shot ... [*opponent hits it*] ... Beauty. [*pause*].

Opponent: Good lord, yes. You've got to have a straight left arm.

Gamesman: Yes. And even that one wasn't as straight as some of the ones you have been hitting.

Opponent: (*Pleased*) Wasn't it? (*Doubtful*) Wasn't it? (*He begins to think about it ...*)

Gamesmanship is not the same as *game theory*, although the end objectives of both are much the same, namely to win. Game theory is the whole art of understanding how to maximise your winnings in a 'game'. The word game is used here in the most general sense and can refer to any walk of life where there are at least two people in competition with each other. It is a much researched aspect of mathematics, of such importance in the military and economic worlds that it has been the speciality of at least two Nobel prize-winners.

Game theory requires you to think not only about what you are doing, but also about what is going through the mind of your opponent. Ted Dexter, the former chairman of the

England cricket selectors, once summed it up nicely when talking about his own strategy for picking a team: 'Always do what your opponents least want you to do.'

Dating

Let's take a simple example of a game. Justin and Tom are in competition with each other. Both of them are keen on a girl called Sally, and both of them want to take her to a party on Saturday. The trouble is that only one of them will succeed, and maybe both will fail. Sally doesn't particularly fancy either of them, and has no preference. She gets back home from school at 4pm, and Justin and Tom can both do one of two things:

- Phone her as soon as she gets home at 4pm; or
- Go to her house and invite her in person.

If they each phone her at 4pm, it is a 50-50 chance which one of them gets through to her first.

Justin lives quite close to her and can visit her in person by 4:15, but Tom can't make it till 4:30 because he has to catch a bus.

Now here is the catch. They both reckon that if they visit her she is 90% likely to accept the date (as long as she hasn't accepted the other lad first). However, they think that she is only 30% likely to accept the date if they phone.

This may all sound complicated, but hey, that's love for you, and calculations at least as complicated as these go through the head of most adolescents.

So what should Justin and Tom do?

You may have your own helpful suggestions on tactics and chat up lines, but to evaluate this game let us instead be more analytical and look at the chances of various outcomes. First look at the situation from Justin's point of view.

If Justin decides to phone and gets through to Sally before Tom does, his chance of success is still fairly low given that she is quite likely to turn him down over the phone. If he gets through second his chance is even lower. What if Justin decides on a personal visit? If Tom has also decided to visit, then since Justin will get to Sally's place fifteen minutes earlier, Justin has a very high chance of getting the date. If, however, Tom decides to phone, Justin now only has a 'fairly high' chance of being accepted because Sally may have already accepted Tom over the phone. All of this can be put into a table, which is known as a *pay-off matrix*: Justin has a choice of phoning (the first column) or visiting (the second column).

	If Justin decides to phone	If Justin decides to visit
...and Tom phones first	Very low chance for Justin	Fairly high chance for Justin
...and Tom phones second	Fairly low chance for Justin	(Not possible)
...and Tom visits	Fairly low chance for Justin	Very high chance for Justin

From Justin's point of view it is always worth visiting, since regardless of what Tom does, Justin always does better if he visits than if he phones. In other words, the right hand column scores higher than the left hand column in each case. This is known as his *dominant strategy.* Notice that there is no need to be exact about figures in this table.[1]

But what is the pay-off table for Tom?

	If Tom decides to phone	If Tom decides to visit
...and Justin phones first	Very low chance for Tom	Fairly high chance for Tom
...and Justin phones second	Fairly low chance for Tom	(Not possible)
...and Justin visits	Fairly low chance for Tom	Extremely low chance for Tom

This is where game theory comes in. Tom's first instinct might be to visit since that's the only way he ever gets a fairly high chance of the date. However, we now know that Justin is going to visit regardless because that's his dominant strategy. This means that Tom will have a 'fairly low' chance if he phones, but only an 'extremely low' chance if he visits. So Tom's best tactic is to telephone Sally.

In this example, game theory gives each player a predictable 'optimal' strategy. Of course just because a player's strategy is optimal, it doesn't mean he is certain to win. In case you were wondering, Sally went to the party with Damian. He has a great motorbike.

1 We haven't quoted exact probabilities, but they can be worked out by using 'very high' as 90%, 'fairly high' as 63%, 'fairly low' as 30%, 'very low' as 21% and 'extremely low' as 9%.

Paper scissors stone

Not all games produce such a definite strategy for each player. For example, take the well known parlour game 'Paper, scissors, stone'. In this game two players each put one of their hands behind their backs and form either an open hand representing paper, a closed hand representing stone or two fingers spread out representing scissors. Then, after the count of three, they simultaneously show each other their hand. If the hands show the same, for example scissors and scissors, then it counts as a draw, and if not then stone beats scissors, scissors beats paper and paper beats stone.

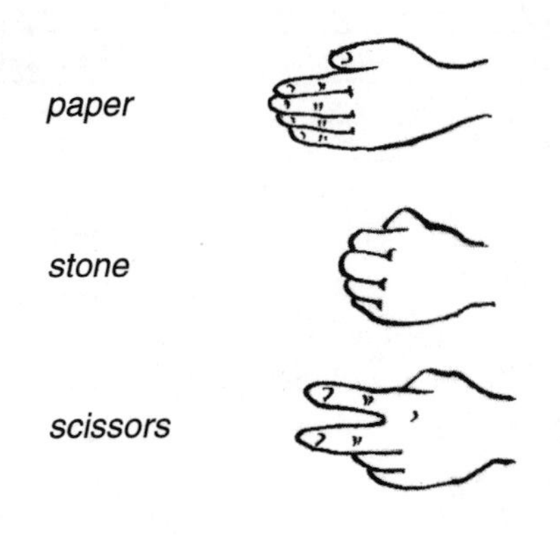

It is possible to represent this game using a pay-off matrix. If a player wins a round they get one point, and if they tie they get no points.
Sid has three possible strategies, which are of course to play paper, play scissors or play stone. Doris has the same three strategies available to her. The table shows the results of each possible play.

	Sid plays paper	Sid plays scissors	Sid plays stone
...and Doris plays paper	Draw	Sid wins	Doris wins
...and Doris plays scissors	Doris wins	Draw	Sid wins
..and Doris plays stone	Sid wins	Doris wins	Draw

This game is very different from the dating game played by Justin and Tom. There is no dominant strategy for Sid or Doris to either maximise wins or minimise losses. Just think of the game from Sid's viewpoint: whichever column Sid chooses he will either win, lose or draw depending upon which row Doris chooses. Similarly for Doris: whichever row she chooses she will either win, lose or draw depending upon which column Sid chooses.

However, if Sid could predict what Doris would do next (or vice versa) then one of them would definitely be in trouble. In paper, scissors, stone there is always a perfect strategy if you know what your opponent is going to play. For example, if Sid knew that Doris was going to play scissors then he would always play stone. But of course Doris would be stupid to continue playing scissors in this case.

In fact for this game the best possible strategy is to randomise your play so that no clues can possibly be given to your opponent about the way you are thinking. As soon as one player's strategy becomes in any way predictable the other player has an advantage, which is what makes the game fun in real life – we try to second guess what our opponent is thinking.

Game theory really comes into its own when the strategy of your opponent directly affects your own decisions. In some games, the interaction between the competitors is almost non-existent. There is no strategy in Snakes and Ladders which is ruled entirely by dice. However, in bridge, football or cricket, out-thinking your opponent is a vital element, and game theory is highly relevant.

Advertising game

Game theory is particularly important in business, and it can lead to some rather paradoxical situations. In a competitive market you will often have two companies producing almost identical products, but desperately trying to do better than each other. With some products such as washing powder or cat food, the total size of the market is fairly constant, and only the market shares can change. (The market is like a cake, with companies competing to get the biggest piece.)

One way of attracting more people to buy your product is to advertise it on television. Let's imagine there are only two brands of toothpaste, 'Dentokleen' and 'Whitefresh'. At the moment, both brands make $2 million profit per year without any advertising. However, the marketing directors in each company know that the other may be about to start advertising. Advertising costs a lot of money, but if you advertise and your competitor doesn't then you stand to make a killing.

In this example we'll suppose that if one company advertises and the other doesn't, the result is that the latter's profits are wiped out. However, if both companies advertise, the effect is simply to cancel each other out, with neither company making any additional profit, and each losing $1 million through advertising costs.

All of this is easier to understand when it is put in a pay-off matrix:

	Dentokleen advertises	Dentokleen doesn't advertise
Whitefresh advertises	Both make $1 million profit	$3 million profit for Whitefresh Zero profit for Dentokleen
Whitefresh doesn't advertise	$3 million profit for Dentokleen zero profit for Whitefresh	Both make $2 million profit

So what is the thinking of the Whitefresh marketing director? He will look at the columns in the matrix and say: 'If Dentokleen advertises, I end up with either $1 million profit if I advertise, or zero if I don't. If Dentokleen doesn't advertise, then my profit is $3 million if I advertise or $2 million if I don't. So whatever Dentokleen does, I am always better off if I advertise'.

The Dentokleen marketing director looks at the rows in the matrix and comes to the same conclusion. So they both decide to advertise.

And yet ... now their profits have fallen by $1 million, whereas if they had both decided not to advertise, they would each have remained at their original $2 million profit. This is extraordinary. Using perfectly sound logic, they have both actually ended up worse off. So who has benefited? Not the general public. Indeed they may end up worse off as well, because with profits down, Dentokleen and Whitefresh may have to put their prices up. The only beneficiary has been the advertising industry, which has gained an extra $2 million in fees.

Supermarket tactic which backfired

In 1996, the British supermarkets began a loyalty card war. Each supermarket reasoned that if they introduced a loyalty card and others didn't, they would increase their own share of the market. Unfortunately, once one supermarket had introduced a card, the others were obliged to do so as well. The result of this game was that the supermarkets gained little from their competitors and spent a lot on cards and customer discounts. It was a classic illustration of why companies would sometimes prefer to operate in a 'cartel'.

Fair play

The toothpaste paradox occurred because the companies were competing with each other and therefore they would not sit down together to collaborate.[2] Furthermore, the free market system actively discourages them from doing so. Just imagine if there really were

2. This is similar to a mathematical paradox known as 'The Prisoner's Dilemma' thanks to a famous example in which two prisoners both confess to a crime and end up with long sentences because they were unable to consult each other.

only the two suppliers of toothpaste. It might be in their interests to put their prices up together so that they would both make more profits, while the people who buy toothpaste would have no choice but to pay the extra. This kind of cartel can make the 'free market' game unfair, which is one reason why, just as in a game of football, there is a need for rules and a referee; in this case the Office of Fair Trading.

Another example of where there are conflicting interests between the individual and the whole group arises on motorways. Imagine the situation where you are warned that there will be a lane closure one mile ahead, reducing three lanes of traffic to two. You find yourself in the middle lane crawling along slowly, but a few selfish individuals go cruising along the outside and cut in a long way ahead of you. Clearly if you want to move faster in the queue than the other cars, it is in your interest to play dirty like this. Yet if everybody played fair and manoeuvred themselves into two lanes when the first warning appeared, the traffic would actually flow much faster. It is because some drivers act in their own interests that everybody is worse off.

These last two examples described situations where people are not playing a strategy which is optimal overall because there is no opportunity for discussion and co-operation. Curiously, sometimes there are games where all the players will actually *agree* that the best strategy is not the one that produces the maximum payout.

Take the game of insurance. Most of us take out insurance. We pay money into a pot, which in this case is the insurance companies' accounts. Nobody except the fraudster plays this game with the idea of making a profit. Indeed in the long term it would be impossible for everyone who is insured to make a profit. The insurance companies cannot pay out more money than they receive, since they need a margin in order to pay for managing their business and paying their shareholders. The reason why we play this game, where we are guaranteed to lose out overall, is that we prefer the certainty of losing a bit to the possibility of losing a lot. (It is similar to the lottery game, where we are happy to lose a little money if it gives us the possibility of winning a lot.)

Games where everybody loses

Finally, there are some games which would never be played if only the participants could step back and be completely rational. Legal action and employment action are common examples of games with outcomes where both parties seem to lose. Often the reason for going to court or going on strike is said to be a matter of 'principle', but principles can be extremely expensive.

Suppose that United Bottlewashers' 5,000 employees are fed up with their pay. They put in a bid for a pay rise of 10 per cent . The management makes an offer of 2 per cent. Both sides say they won't budge. We all know from countless previous examples that in the end the two sides will settle somewhere in the middle. By the time all the factors have been taken into account the settlement will probably be somewhere between 5 per cent and 7 per cent. Sadly the two sides refuse to budge, the negotiations go nowhere, and the workforce goes on strike. After a month of acrimony United Bottlewashers settle with the unions at 6 per cent. Both sides hail it as a victory.

Millions of pounds are diverted to the wages budget, but the price of this is high to everyone. The company has lost short term sales and several regular customers, as well as money from its other budgets that has been switched into labour costs. The workforce meanwhile may have gained more salary, but it can take a year or more to earn the money to compensate for income lost during the strike. Some employees may lose their jobs altogether in productivity deals.

It could well be that the medium term cost to the company of a strike far outweighs the saving made in not giving the full pay rise, while going on strike means that the workforce can end up earning less money as a whole than if they had accepted the original offer from management. Everyone seems to have lost. However, like the advertising game we looked at earlier, without consultation and co-operation, both sides can find themselves forced to adopt the 'lose-lose' strategy.

If game theory can teach us anything, it is that there are some games that should be avoided altogether.

12

WHO'S THE BEST IN THE WORLD?

The mathematics behind sports rankings.

From early childhood onwards there seems to be a basic urge to rank people in order. Every child learns his or her worth from the dreaded line-ups in the playground when two captains pick their teams. This trait of ranking in order goes on into adulthood (although it usually seems to be of more concern to men than women). We like to know who is first and who is last, who is improving and who is getting worse. It doesn't seem to matter that ranking can mean sweeping simplifications of reality. There is something very seductive and convincing about putting one name at the top of a list.

Nowhere is ranking more important to the general public than in sport. Rankings make good headlines, and help to meet the public's demand for an answer to the question 'who's the best in the world?' Some rankings, like those in tennis, can determine whether a player appears in a tournament or not. Others, like those in international football and cricket, are mainly there for the interest of the spectator.

Anybody can write down their personal ranking of favourite players, but feelings about sports stars run so high that it would be quite unacceptable for official rankings to be decided by individuals. The Eurovision Song Contest can get away with it, but no sports fan would trust the judgment or independence of panels of experts unless they produced exactly the top ten that the fan believed was right. This is why sport puts its faith in math to calculate the rankings. After all, mathematics is precise, logical and objective, so why not?

Unfortunately, even mathematical rankings cause controversy. It seems that very often in rankings, the 'best' players are not top. How can this be?

What happened before sports rankings?

Thirty years ago there were no official world rankings. Competitors would often be selected for tournaments purely on the personal whims of the organising panel. Alternatively, the total prize money won over a period of time would be the qualification. Both systems were the subject of controversy. A player could be excluded simply because his face didn't fit, and money rankings could be distorted if a wealthy country decided to offer prizes that were far higher than the importance of their tournament merited.

In 1973 the Association of Tennis Professionals (the ATP) had had enough of the old subjective methods. They devised a points system so that players could be compared 'objectively'. Mathematical modelling had made its first serious entry into the world of sports rankings, and since then practically every team and individual sport has followed suit.

The First Tennis Rankings

23 AUG 1973	
1	Nastase
2	Orantes
3	S. Smith
4	Ashe
5	Laver
6	Rosewall
7	Newcombe
8	Panatta
9	Okker
10	Connors

Why can't sports rankings be simple?

The whole point of a ranking is that it should be fair. It also helps if the workings can be understood by the general public. Complicated mathematical formulae arouse suspicion because sports fans can't follow the reasoning behind the results. Unfortunately fairness and simplicity don't always go hand in hand.

Take two simple ways of producing a sports ranking.

1. Add up the points for every tournament.
In this ranking method, each tournament has points available, and players or teams gain ranking points by playing in as many tournaments as they can. It is the easiest form of ranking, and is ideal for football leagues, for example, where each team has to play the same number of games. It also works in Formula 1 racing.

But in individual sports like tennis or golf there is a problem, because there are more tournaments than players can physically play. In sports like tennis, players are prone to injuries, and a ranking which adds up all the points might encourage players who should be resting an injury to compete instead. Rankings of this sort become a measure of who is the fittest as much as of who is the best. This is even true in baseball and basketball, where the team with the biggest squad can have an advantage at the end of the season.

2. The law of averages
One way of avoiding a ranking based on fitness is to produce a ranking of the competitors based on the average of their performances. Baseball batting averages work this way. Suppose a coach has two players on his team who play the same position in the outfield and have about equal fielding ability. He's been playing one of them about twice as much as the other, and he wants to evaluate whether this is the right strategy. One thing he can do is to calculate their batting averages. Is the player with 25 hits in 125 at bats hitting better or worse than the player with 16 hits in 60 at bats? To get the average, he takes the number of hits and divides by the number of at bats. In this way, he calculates that the player who has 28 hits in 125 at bats is hitting .224, and the player with 16 hits in 60 at bats is hitting .266, appreciably better. The coach might want to consider giving the second player more play time. But note that this is only a crude measure of the players' comparative contributions to the team's success so far in the season. The first player may have had fewer hits on the average, but if he's batted in more runs with those hits, maybe because he's hit more home runs or doubles and triples, he may actually be the more valuable batter.

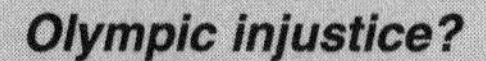

Olympic injustice?

In the Olympic Games, countries have for years been ranked by newspapers according to the number of gold medals they have won. This simple league table is widely accepted, but it ignores the significance of the events in which the golds were won, and arguably puts too much emphasis on gold and not enough on silver – as Great Britain knows to its cost.
In the 1996 Olympics, Britain (1 gold, 8 silvers) appeared behind Algeria (2 golds, 0 silvers) in the league table. Britain had a bad games, but were they really worse than Algeria? 4 points for a gold, 2 for a silver and 1 for a bronze might produce a fairer ranking table.

If an average points system was used in tennis and a player was not obliged to take part in tournaments, he might refuse to play simply because to do so would damage his ranking. For example, if a player has 10,000 points in 10 tournaments (average of 1,000) then if he plays in an eleventh tournament and scores 0 points his average drops to 10,000 / 11 = 909 points. He might not want to take this risk.

Both of the above systems are also distorted if some tournaments attract stronger players than others. Winning a minor snooker tournament in Bognor Regis should obviously not be worth the same number of points as winning the world championship. If it was, then all sorts of unknown players would top the lists.

It is basic problems like these that have led most sports to devise more complicated ways of ranking. The basic principles in most of these rankings are the same:

- More points are available for winning the major 'tough' tournaments;
- There is usually a combination of a player's average performance (to reward excellence) and his cumulative total (to reward the hard workers).
- Most rankings take account of previous years' performances, although what a player did the previous year counts for less than what he does in the current year.

The exact ways they do this vary from sport to sport. For example the Sony rankings for golf are based on a simple average, but with a clever adjustment to handle those players who do not play enough tournaments. The ranking is calculated by dividing the player's points total by the number of tournaments he competes in. However, if the player competes in fewer than 10 tournaments, his total is still divided by 10. For example (and these figures have been made up):

Faldo	60 points in 12 tournaments
	Ranking average = 60/12 = 5.0
Ballesteros	40 points in 8 tournaments
	Ranking average = 40/10 = 4.0

Even so, this system will tend to favour somebody who plays only 10 tournaments over somebody playing 30, since it is easier to sustain good performances over a small number of tournaments than over a large number.

Tennis and football rankings pick out only the best performances of the year (the best 14 and the best 8 respectively). This tends to slightly favour those who play large numbers of tournaments, since they have more good results to choose from.

Another problem in ranking arises at the end of the season in some sports, when teams vying for playoff spots have played different sets of competitors. For example, in college football, disputes about rankings often result when a team with a higher number of wins has played against an easier lineup of competitors than another team with one or two fewer wins that has played a very difficult lineup of competitors.

Already it is beginning to look as though no sport has found the perfect answer to ranking players using points. But even if the conflict between averages and cumulative points were to be resolved, mathematical rankings in sport would still lead to results that appear to have anomalies.

A game of two halves

How easy is it to draw the wrong conclusions from a table showing a ranking? Here are the results of a league, in which names of teams have been replaced by letters. There are ten teams in the league, and this is the end of season table so each team played each other twice, once at home and once away. There are three points for a win and one point for a tie.

	Played	Won	Drawn	Lost	Points
A	18	11	2	5	35
B	18	9	4	5	31
C	18	9	3	6	30
D	18	8	3	7	27
E	18	7	5	6	26
F	18	7	3	8	24
G	18	6	5	7	23
H	18	5	6	7	21
I	18	3	8	7	17
J	18	3	5	10	14

Which team's manager would you sack? If A plays a match against J next week, who will win?
A safe answer is to sack team J's manager and to put money on A to beat J next week. Yet in this case the conclusions are completely wrong, because ...
These teams were not sports teams, they were people tossing coins against each other. The results are quite genuine. Depending on what came up on the coins, the result was a win, draw or loss for the teams. Each team played to the same rules, so they all had an identical chance of winning. There was absolutely nothing that the managers could do to influence the results, and therefore praising or criticising the managers would be meaningless. And if Team A plays against Team J in the next match of coin tossing, the teams have an equal chance of winning. After all, if you toss a coin and it comes up heads five times, the sixth toss is just as likely to be a tail as a head.
What is odd, however, is that the table looks just like an ordinary table of soccer teams. Does this mean that a soccer league is little more than a group of teams tossing a coin every week with the managers biting their nails in case the coins don't come up in their favour? In real soccer, the coins are probably biased towards some teams, but luck certainly plays a part and is one reason why you shouldn't read too much into any sort of sports ranking.

Quirks and anomalies

There are three common types of anomaly in sports rankings which are unavoidable and can lead to confusion or ridicule:

1. Player moves up the rankings despite performing badly, or not playing

In 1992, Stefan Edberg went top of the ATP rankings at the end of the week when he suffered the worst defeat of his career, losing to Robbie Weiss, who was ranked 289. This was an extreme example of the sort of odd result that can happen in nearly all sports rankings.

In tennis rankings, a player's points for the current year's tournament replace the points he earned in the tournament the previous year. Edberg had performed poorly in the tournament the previous year, so even his mediocre performance in 1992 was enough to gain him points.

In tennis, the round of the tournament you progress to is what counts, not the ranking of the player who beats you. If Edberg had been knocked out by Andre Agassi, he wouldn't have got any more credit.

The explanation for Edberg's move up the rankings was perfectly logical, but without knowing the simple principle involved, a headline along the lines of 'Edberg goes top after ignominious defeat' would make the ranking system seem ludicrous.

A related phenomenon sometimes happens at the end of the baseball season, if several teams are vying for a spot in the playoffs. In some cases an inferior team may secure a spot in the playoffs not by winning any additional games, but simply because a better team loses a game or two.

2. Rankings treat stars and ordinary players the same way – but the public don't.

Some sportsmen attract more headlines than their performances merit: Paul Gascoigne, Andre Agassi and Brian Lara have been examples in recent times. They gain their reputations for their moments of genius, for their looks, for scandal or for human frailty (sometimes all four). But their prominence in the public eye means that people expect them

"The boys played 110%"

How much better does a team have to be than its opponents if it wants to win? Sometimes the difference needs only to be very small. If a British tennis player is the same strength as an opponent but improves his service by 10% because of home support at Wimbledon, this will be enough to win him the match. A football club only needs to be about 20% more likely to win its games than other clubs in the league to have a strong chance of winning the league most years. If the players really could play '110%', maybe it would be enough to turn a middle place in the league to the top place.

to be high up in the sports rankings at all times. Media attention is not something that can be conveniently translated into points.

3. Computer fails to capture the subtlety and magic of a sporting occasion
Perhaps this factor is the most important reason why it is hard to rate sportsmen using statistics alone. There is no mathematical formula for human emotion, but that is what many sporting performances are remembered for. American gymnast Kerri Strug helped her team to the Olympic gold medal in gymnastics after twisting her ankle. Greg Norman's golf fell apart in the final round of the US Masters in 1996 which allowed Nick Faldo to win. And in 1998, Mark McGwire and Sammy Sosa became sporting legends when they both beat Roger Maris's longstanding home run record.

The hostility of a crowd, luck, tension and sudden drama are very hard to quantify, and as a result are ignored by rankings, but since these are real factors that contribute to a sportsman's greatness there will always be a discrepancy between hard figures and public perceptions.

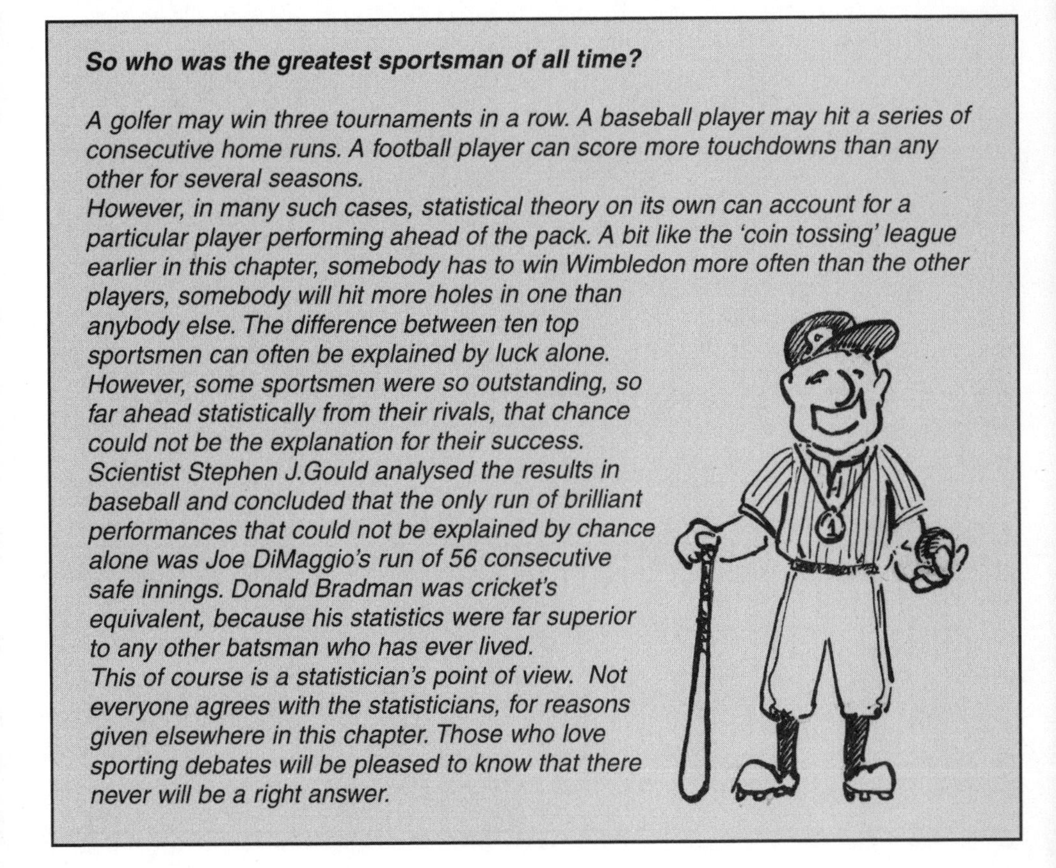

So who was the greatest sportsman of all time?

A golfer may win three tournaments in a row. A baseball player may hit a series of consecutive home runs. A football player can score more touchdowns than any other for several seasons.
However, in many such cases, statistical theory on its own can account for a particular player performing ahead of the pack. A bit like the 'coin tossing' league earlier in this chapter, somebody has to win Wimbledon more often than the other players, somebody will hit more holes in one than anybody else. The difference between ten top sportsmen can often be explained by luck alone.
However, some sportsmen were so outstanding, so far ahead statistically from their rivals, that chance could not be the explanation for their success. Scientist Stephen J.Gould analysed the results in baseball and concluded that the only run of brilliant performances that could not be explained by chance alone was Joe DiMaggio's run of 56 consecutive safe innings. Donald Bradman was cricket's equivalent, because his statistics were far superior to any other batsman who has ever lived.
This of course is a statistician's point of view. Not everyone agrees with the statisticians, for reasons given elsewhere in this chapter. Those who love sporting debates will be pleased to know that there never will be a right answer.

14

WHAT HAPPENED TO CHAPTER 13?

Can bad luck be explained?

Toast always lands butter side down. It always rains on bank holidays. You never win the lottery, but other people you know seem to.... Do you ever get the impression that you were born unlucky? Even the most rational person can be convinced at times that there is a force out there making mishaps occur at the worst possible time. We all like to believe that Murphy's Law is true ('if it can go wrong, it will').

Part of the explanation for bad luck is mathematical, but part is psychological. Indeed there is a very close connection between people's perception of bad luck and interesting coincidences (see Chapter 6).

For example, take the belief that 'bad things always happen in threes' (just like buses ...!) This popular notion would be unlikely to stand the scrutiny of any scientific study, but it must have some basis in experience, otherwise the phrase would never have arisen in the first place. What might be the rational explanation?

The first question is 'what is bad'?

Some things are only marginally bad, for example the train arriving five minutes late. Some are extremely bad, such as failing an exam or being sacked. So badness is much better represented as being on a spectrum rather than something which is there or not there.

A particular event may only be a misfortune because of the circumstances around it. The train arriving five minutes late is a neutral event if you are in no hurry and reading an interesting newspaper article while you wait. It is bad if you are late for an important meeting.

When it comes to bad things happening in threes, what may be most important of all is the duration and memorability of the first event. Take a burst pipe while you are away on holiday, for example. It may take less than an hour to flood the house, but this one bad event can remain alive and kicking for many months, with the cleaning up operation and the debate with your insurers acting as constant reminders of the original event.

The longer the first bad event sticks in the front of your mind, the more opportunities you will have to experience two more bad events. A month later someone bumps the back of your car and a week after that you lose your wedding ring. The mind which is already on a low from the first event will quickly leap to connect the subsequent misfortunes as part of the series. It wouldn't matter that there could be a two month time scale over which everything happened. By the time you have recovered from the water damage you are actively looking out for the next disaster. The timescale has been extended as long as is necessary to confirm the original prophecy.

As with coincidences, in bad luck there is a tendency to look for the examples which confirm the theory, and ignore those which don't (because they are less interesting). Single bad events happen all the time. That alone should be enough to disprove the theory. Bad things also come in twos. But it is more likely that a friend will tell you 'three bad things have happened to me, isn't that typical' than 'only two bad things have happened to me, which just proves that the theory doesn't work'. After all, the latter is tempting fate!

There is, however, at least one rational reason why bad events might cluster together. It is related to probability and independence (p.49). Unlucky events are not always independent of each other. Anybody who is made redundant is bound to suffer some depression. That will lower the body's defences, making the person vulnerable to illness, and also making them less alert and responsive (so they may be more likely to drop a precious vase, for example). So while the probability of being made redundant on any particular day and the probability of being sick may both be small, the chance of both occurring is almost certainly higher than the product of the two probabilities.

Map reading misfortunes

So much for the general incidents of bad luck which crop up in life. Let's get on to a specific one that everyone has encountered.

You are off to visit a friend who lives at the other end of the city. You look up the road in the street atlas, and discover that it is right on the edge of the page. This means that finding the precise route becomes a chore of flicking backwards and forwards from one page to the next. Either the road is half on one page and half on the other, or it's spread across the fold in the middle of the book. And if it's an ordnance survey map, then your destination is at just the point where you folded the map over.

It doesn't seem fair. After all a map only has a tiny bit of 'edge' but plenty of 'middle' in which your destination could be situated. Or has it? In fact the chance of picking a destination which is close to the edge of the map is a lot higher than you might expect.

Take a look at the map in the diagram.

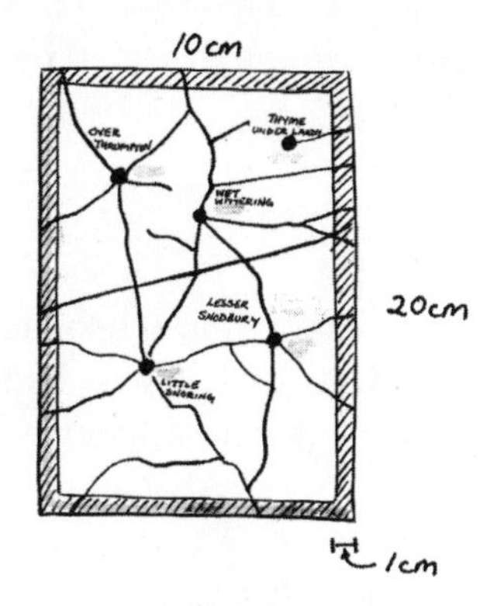

Each page is 10 cm by 20 cm – a total map area of 200 cm ^{2}.

The shaded area is 28% of the whole area.

You will have a problem if your destination is anywhere in the shaded area marked on the map. This shaded area is just 1cm into the page all the way around. It looks insignificant. However, the shaded area adds up to 56 cm^2. That represents 28 per cent of the area of the whole page of the map, which means that any specific point that you are seeking on this map has a 28 per cent chance (that's nearly one in three) of being in an awkward position within 1cm of the edge of the page. And if you regard being within 2cm of the edge of the page as being awkward, the chance of ill-fortune climbs to 47 per cent. In other words, you might expect this misfortune to occur on almost every other journey.

As in most bad luck stories, you forget about the number of times the road doesn't land awkwardly and remember the times it does, and in this case the chance of a bad result is so high that before long you are bound to be cursing your misfortune, or the map's printer, or both. This, incidentally, is why many modern road maps allow significant overlaps between adjacent map pages. In a good road atlas, at least 30 per cent of the page is duplicated elsewhere.

The lights are always red when I'm in a hurry

One of the best examples of selective memory where an unfair comparison is made between good and bad is in the relative frequency of red and green lights on a journey. For once, the perception of 'I always seem to get red lights when I'm in a hurry' is true and verifiable. To simplify the situation, think of a traffic light as being like tossing a coin, with a 50% chance of being red, and 50% of being green. (In fact most traffic lights spend more time on red). If you encounter six traffic lights on a journey, then you are no more likely to escape a red light than you are to toss six consecutive heads, the chance of which is 1 in 64.

Red lights come up just as often when the driver is not in a hurry; it's just that the disadvantage of the red light is considerably less if time is not critical. The false part of the perception is that red lights happen more than green lights. The reason for this is simply that a driver has more time to think about a red light than a green light, because while the latter is gone in seconds – and indeed is an experience no different from just driving along the open road – the red light forces a change of behaviour, a moment of exertion and stress, and then a deprivation of freedom for a minute or so. Red lights stick in the mind, while green lights are instantly forgotten.

Everyone else wins the lottery, so why don't I?

Finally, of course, there is the lottery. Adam comes home from school and says 'Jason's aunt just won $500!'. 'Well the family of somebody at our school won $1000,' says his sister Melanie. 'That's amazing,' says dad, 'a guy at work was telling me yesterday that a friend of his had had a big win too.' The family can't believe that all their friends appear to have been lucky and not them.

The fallacy here, of course, is that none of the people that they reported as having won in the lottery were their friends. In fact there is little that is surprising about this story. Melanie has reported the biggest win, but how many people could her story have come from? There could be a thousand people at her school, each of whom has perhaps ten people who might count as family. That makes 10,000 possible sources from which the story could have come. A decent lottery win may have a shelf life of four weeks as a story.

> ***Unlucky 13***
>
> *13 is a notoriously unlucky number, though it is unclear where the superstition arose. Superstitious architects avoid 13th floors in tower blocks, and superstitious authors might even choose to avoid a 13th chapter. The most famous unlucky event associated with 13 was the Last Supper. Friday 13th is the unluckiest day to hold an event, and it just so happens that the 13th day of a month is more likely to occur on a Friday than on any other day. This is a statistical fluke which is a result of the cycle of days in the Gregorian calendar.*

If in the last four weeks those 10,000 people bought a total of 10,000 tickets between them, then it becomes really quite likely that one of them would achieve a $1000 win. Stories of big wins spread rapidly, but non-winning tickets are quietly forgotten. All 9,999 of them. Adam's story is just as easily explained, and dad's is so vague that it is unclear whether the win was made last week, last month or last year.

Maths has a lot to say about luck and misfortune. However, there is no doubt that whatever rational reasons lie behind it, some people seem to attract more luck than others. Emperor Napoleon used to have a policy of promoting generals who were 'lucky'. Superstitious nonsense? Not necessarily. Napoleon's shrewd military brain would have told him that whatever made his generals lucky in the past would probably do so again. Or, as the golfer Arnold Palmer once said: 'the more I practise, the luckier I get'.

Bad luck turns good – a story of how bad luck can be exploited.

"When my husband Harold was young, he was unfortunate enough to have an argument with a witch," said Margaret. "She was so angry that she put a curse on him – he was doomed to be unlucky for the rest of his life. He's forever late because the train was cancelled; if there's a cold going round you can bet he will catch it, but worst of all he is addicted to gambling and of course he is a serious loser. Every time he goes to the casino he loses a fortune."

"That must be terrible, Margaret. I'm amazed you have stayed married to him."

"Oh, it's the best thing that could have happened to me. I've made my million, and I can more than cover Harold's losses!"

The reason for her wealth?

Margaret always accompanies Harold to the casino. Whatever Harold places his money on, Margaret puts double on the opposite!

15

WHODUNNIT?

Everyday logic, from murder mysteries to political statistics.

There is a Sherlock Holmes story called *Silver Blaze* about a corrupt horse trainer who wants to nobble his own horse. In it, Holmes makes one of his most famous pieces of deduction. Inspector Gregory asks Holmes if there is any evidence he wishes to draw to the inspector's attention.

> *'To the curious incident of the dog in the night time,' [replied Holmes].*
> *'But the dog did nothing in the night time,' [said the inspector]*
> *'That was the curious incident,' remarked Sherlock Holmes.*

Holmes reasoned that the dog didn't bark because he recognised the intruder, hence proving that the intruder must have been the dog's owner.

This example of inference demonstrates that what *isn't* said can sometimes be as helpful as what *is* said. Politicians give away information all the time when they say 'no comment'. After all, if they had something positive to say, surely they would be only too keen to comment.

Politicians give away information in other ways, too. Suppose that the President announces: 'I am delighted to say that unemployment has fallen in three of the last four months'. What can you deduce about what happened to unemployment four months ago and five months ago? At first glance, nothing at all.

But remember that politicians are masters of presenting information in as favourable a way as possible. Suppose unemployment *rose* four months ago. This would mean that it had fallen in each of the last three months, so if this had been the case the President would surely have made the more impressive statement 'unemployment has fallen in each of the last *three* months'.

Similarly, if unemployment had fallen five months ago, he would have announced proudly that 'unemployment has fallen in four of the last five months'.

This trick is used all the time in the presentation of statistics, and it is fun looking out for it. In the coincidence chapter (page 48) is the statement 'three of the first five Presidents of the USA died on 4 July'. The fifth President, Monroe, was one of the five, otherwise the coincidence would be presented as even more remarkable 'three of the first *four* Presidents'.

Party hats game

Here is a little deduction game which makes an interesting experiment. You need three paper hats, two of which are the same colour (let's say two reds and one blue). You also need two volunteers to sit facing each other.

Show them the three hats, then get them to close their eyes while you put a hat on each of them. When they open their eyes, they have to deduce the colour of their own hat simply by watching the reaction of the other person.

What they don't know is that you have given them both a red hat. Many adults are

unable to make any deduction at all (they just guess). But the following reasoning by either volunteer would lead to an accurate deduction.

"Suppose I have a blue hat on. The other person knows there is only one blue hat, so he will immediately realise that he must be wearing a red hat. But he hasn't made this simple deduction, so I can only assume that I am wearing a red hat, not a blue one."

There is obviously a lot that we can learn from Sherlock Holmes. He is still one of the most popular fictional detectives, fascinating not only for the nature of crimes that he solved but also for his own personality. Holmes was renowned for being able to stick purely to the facts and to apply pure reasoning without allowing his emotions to get in the way. He was one of the greatest role models for any logical thinker, although it appears that Holmes never told jokes, so he wouldn't exactly have been a bundle of laughs at parties.

But what does Sherlock Holmes have to do with math? Truth and lies, implications and deductions, consistency and inconsistency are part of everyone's daily life. We use them and live by them without any consideration for math at all. What math does is give logic a very precise language and notation which helps to ensure that the logic is watertight. Without watertight logic, it can be easy to come to the wrong conclusion. Take a famous example from *Alice in Wonderland.*

'Then you should say what you mean,' the March Hare went on.
'I do, Alice hastily replied; 'at least – at least, I mean what I say – that's the same thing, you know.'
'Not the same thing a bit!' said the Hatter. 'Why, you might just as well say that "I see what I eat' is the same thing as 'I eat what I see!"'.

Deductions – right or wrong

Deduction applies to more than just the world of crime. In fact almost any conversation has built into it deductions and implications, often signalled by the word 'therefore'. But how often are these deductions false?

Children find the following argument great fun. 'When it is windy, the trees wave their branches around. Therefore wind is caused by trees waving their branches.' You may smile, but how do you prove to a child that this statement is not true? You need to find a counter-example, such as a desert where it is windy but there are no trees, or even better, a tree waving its branches where there is no wind. The latter is possible, but it takes a lot of effort to shake a tree!

All zebras are striped animals, but does this mean that all striped animals are zebras? Of course not. It is similar to Alice's error. 'I mean what I say' doesn't necessarily mean 'I say what I mean'. (Actually working out the difference between these two statements is quite tricky).

Could you fall for a logical error like the zebra? Try this.

A man deals out four cards from a pack in which each card has a shape on one side and a pattern on the other side. The man then claims: 'Any card on this table with a triangle on one side always has stripes on the other side'.

Which cards do you need to turn over if you want to be certain that his claim is true?

You may want to decide on your answer before you read on.

A most common response is to turn over the triangle card and the striped card. However, the correct answer is that the triangle card and the spotted card need to be turned over. If you turn over the spotted card and it has a triangle on the back, the man's statement was false. Turning over the striped card to find a square or the square card to find stripes proves nothing.

The confusion here is that the statement 'all triangle cards are striped is not the same as 'all striped cards are triangles'. It is the zebra fallacy by another name! A convenient way to demonstrate this is to use what is known as a *Venn diagram.* This way of representing logic was popularised by John Venn in the 19th century.

Suppose that all of the zebras in the world and all of the striped animals in the world have been rounded up. The zebras are now placed in a big circular cage. Anything in the cage is a zebra. Anything outside the cage is not a zebra.

Now you need a cage to include all striped animals. Tigers, ring-tailed lemurs and lots more besides will be inside the cage, but so will all the zebras. (Let's ignore any zebras which manage to have grown up without stripes!)

The way to do this is to build the striped animals cage around the zebra cage. Zebras are a special case in the *set* of striped animals. A set is a collection of things labelled by what they have in common, in this case stripiness.

This is all pretty straightforward, but what it does is allow a simple visual method for demonstrating fallacious arguments.

Zebra fallacies feature regularly in legal work. For example in one case, a man was suing his company for having caused his deafness. It was agreed by both sides that 'prolonged exposure to loud machinery in the factory led to hearing loss.' However, the plaintiff's argument was that this *proved* that his damaged hearing was due to prolonged exposure to loud machinery. This may have been true, but there are many more causes for damaged hearing than loud machinery – a defective gene for example. To further complicate matters, having a certain defective gene may only sometimes lead to hearing loss.

This simple example can be put into a Venn diagram showing three sets: those who have damaged hearing, those who have had prolonged exposure to loud machines, and those who have the defective hearing gene.

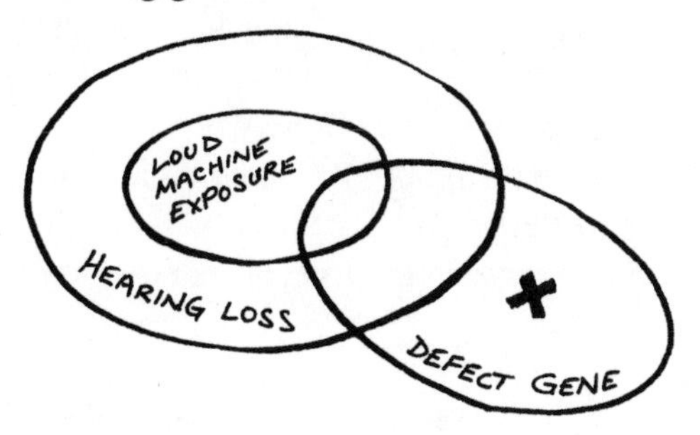

Notice that the 'loud machinery' set is inside the 'hearing loss' set. This represents the original statement that everyone who has prolonged exposure to loud machines will experience hearing loss. However, it is clear from the diagram that not everyone with hearing loss has been exposed to loud noise. Some people are outside the machinery set but inside the hearing loss set.

The set of people with the defective gene overlaps both of the other sets. Some with the defective gene have had too much noise and have suffered hearing loss. Some have simply suffered hearing loss. Those in the area labelled X still have good hearing.

Mathematicians and logicians use special notation to represent regions within a Venn diagram, but it's not important to know it. The picture says it all.

Children, logic and NOT

Children can understand some complex logic at a very early age. For example, a three year old child can understand the statement 'If you don't put your coat on, you can't go outside'. This may lead to foot stomping and tears, or to immediate compliance, or possibly both.

It is interesting that someone so young can understand that two negatives in language make a positive, but will probably take another few years before being able to make the same interpretation in simple arithmetic. The NOT function is one that a child masters particularly quickly, probably because it features such a lot in early life. (Do NOT make that noise, do NOT touch that plug, AND do NOT do that to the dog ...)

Writers are usually discouraged from using double negatives because they can be confusing. Even worse are treble or even quadruple negatives. On a recent discussion about TV programmes, one interviewee said: 'I'm not saying that there aren't programmes that are unsuitable for children before nine o'clock ...'

If you are struggling to work out what she meant by this, it is probably because she used three negatives in that statement. They were 'not saying', 'aren't programmes' and 'unsuitable'. One of the quickest ways to make sense of a statement like this is to cancel out two of the negatives. So what the woman was saying was along the lines of:

"I *am* saying that there *are* programmes that are unsuitable for children ...'

Unfortunately it isn't quite as simple as that. If you say '*I am not poor*' does that mean the same as '*I am rich*'? No, not quite. Again, a Venn diagram explains it very quickly. The world could be divided into the rich, the middling and the poor. These are three exclusive sets – nobody can be in two at the same time. Those who are *not* poor are all those who are outside the low income cage in the diagram. But they are not exclusively rich – there are the middlings as well. Hence not poor isn't always rich, although it sometimes is.

This principle of statements always being 'true' or 'not true' (or false), and things being

LOW INCOME $

MIDDLE INCOME $

HIGH INCOME $

'inside' or 'outside', but never both, is fundamental to traditional logic, although we will see later that this does have its limitations. Anyone practising in detection, science or the law relies on it profoundly.

Hence the remark by Sherlock Holmes: 'When you have eliminated the impossible, whatever remains, *however improbable*, must be the truth'.

This thinking is reassuring for those of us who have, for example, left a passport lying somewhere around the house. Once you have searched all the rooms in the house from top to bottom and failed to find it, then the improbable is all that remains. Maybe it *did* fall down a crack in the floorboards. Maybe it *was* eaten by the dog.

Computers and logic gates

We've seen that true and false statements can be represented by being inside or outside a bubble or cage on a Venn diagram. Computers do exactly the same thing, except they represent true and false with numbers instead. True is 1 and False is 0.

Every logical check in a computer that interprets an instruction is known as a *gate* (not to be confused with Bill Gates, although he has been responsible for most of them). If you were somehow able to look into a computer's circuitry with an all-seeing microscope, you would discover that the whole thing is built out of just three simple logical functions:

- NOT
- AND
- OR

Here is a simple NOT statement any child will understand. 'If you scream, I won't read you a story'.

A computer would interpret this as follows:

INPUT *(Scream)*	***OUTPUT*** *(Read story)*
IF INPUT is TRUE then	NOT Read story
IF INPUT is NOT TRUE then	Read story

or as a table:

Input (Scream)		Output (Story)
1	->	0
0	->	1

An AND statement would be. 'If you eat your chicken and eat your sprouts, I will read you a story'. There are two inputs here, chicken and sprouts. Here are the results as a table.

Input 1 (Chicken)	Input 2 (Sprouts)	Output (Story)
1	1	1
1	0	0
0	1	0
0	0	0

So the only input that delivers a story is when chicken and sprouts are both 'TRUE'.

Finally, 'If you wear a hat or use an umbrella you will stay dry' is an example of an OR function.

Input 1 (Hat)	Input 2 (Umbrella)	Output (Dry hair)
1	1	1
1	0	1
0	1	1
0	0	0

And that's computers for you. Everything from calculating the square root of five to landing a spacecraft on Mars is made of AND, NOT and OR gates. Millions of them linked in the correct way (that is the tricky part, as you might have guessed).

Is the human brain also composed purely of AND, NOT and OR gates? This is the big question behind Artificial Intelligence. There are many who think that our brains work in a completely different way, which is why human logic can be fallible but also far more creative than a computer ever will be.

THIS STATEMENT IS FALSE

Is the above statement true or false? If it is true then it is false. But if it is false then it is true! This little paradox has been at the heart of great dissertations on logic and philosophy because it shows that the notion of 'true' and 'false' does not always apply to a statement.

Fuzzy logic

In recent years, computer programmers have begun to realise that one big difference between human and computer logic is that humans don't always think 'yes' or 'no' but sometimes they think 'maybe'.

Somebody may describe the weather by saying 'It is a sunny day'. Surely this statement is absolutely unambiguous? Unfortunately not. Suppose there is one cloud in the sky. Maybe it is still a sunny day. Two clouds? Yes. A thousand clouds? No, now it is a cloudy day. But this means that somewhere between two and one thousand clouds the day moved away from being a sunny day. It didn't happen immediately, but it happened by degrees, although there was some absolute point of cloudiness at which any individual would have suddenly refrained from calling it a sunny day. Nobody can pin down that moment, so there is fuzziness in the statement that it is a sunny day.

People tend to dislike cut off points because they artificially divide things which are quite close to each other. One example is the residents who complain that one half of a street is in a borough with a low council tax while the other half is in a different borough with a high tax. 'Why should I be paying the council £900 when those people opposite only pay them £400 for the same thing?" Ideally, council taxes should be fuzzy.

Fuzziness also applies to statements such as 'Susan looks just like Jill'. How close does Susan's appearance have to be to Jill's before this statement is true. There is no right answer, but a human would be comfortable making the statement and a computer,

EXOR and the light switch in the hall

Most homes have two separate switches that operate the hall light, one by the door downstairs and one upstairs (for instance). This is the commonest example of what is called an Exclusive OR (EXOR) function. This is not the same as an OR function. If either the downstairs switch or the upstairs switch is on then the light is on, but if both switches are on, the light is off.
A computer circuit could be built to behave as an EXOR gate using the AND, NOT and OR gates as follows:

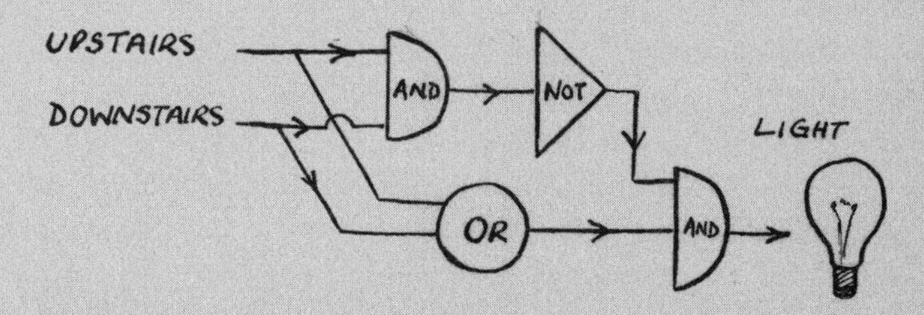

Using 1 to represent 'switch is on' and 0 to represent 'switch is off', this produces the same result on the light as two mechanical switches. To test it you can try each of the four possible input pairs, 0,1 1,0 0,0 and 1,1.

traditionally, would not. This is why programmers have started moving from categorising everything as absolutely True (1) or False (0) to attaching values in between. A half true statement would be 0.5.

Fuzziness fits well with the two professions that appeared at the start of the chapter. Detectives like Sherlock Holmes would always allow a 'maybe' to exist until a culprit had been proved guilty. And any politician knows it is fuzziness that enables them to stay in office. At least, we might say that. They couldn't *possibly* say it.

16

WHY AM I ALWAYS IN TRAFFIC JAMS?

Motorways, escalators and supermarkets all have one thing in common: queues.

John lives at number 10 King Street. A creature of habit, he leaves for work at 7:30am on the dot every morning and within a few minutes either way arrives at the office of United Bottlewashers, in the city centre, for 8:00am.

John's neighbour Brian also works for United Bottlewashers, but he always leaves for work a bit later. What with feeding the cat, leaving a note for the milkman and ironing the shirt he forgot to do the night before, he often gets in his car at about 7:40am and arrives at the office at about 8:30am, by which time John is writing his fourth memo.

John and Brian take the same route, have identical cars and have exactly the same driving habits (the same top speed, the same acceleration etc) yet Brian's journey takes him twenty minutes longer. How can this be?

You will of course realise that this is not a trick question but an example of a daily reality. Many car commuters will recognise the problem of 'if I get out of the house five minutes late it usually costs me half an hour'.

But why does this happen? It is something to do with traffic build up, of course, but what has it got to do with maths? The answer is that traffic belongs to a delightful part of mathematics known as *queueing theory*.

Traffic lights

Imagine that John's and Brian's route takes them on to City Road, which has a single set of traffic lights. Traffic lights in cities are usually programmed to be sensitive to traffic. If, over 30 seconds say, no cars drive over the sensors in front of the lights, then the lights are likely to change to red. However, in rush hour with traffic constantly running over the sensors, the lights stay green at a pre-programmed level. In City Road, the traffic light sequence happens to be 20 seconds of green followed by 40 seconds of red, and that one phase of green is enough time to let ten cars get through. This means that on average ten cars a minute are getting through the City Road lights. This is what is known as the 'service rate' of the lights.

The number of people leaving home in the early hours probably starts as a trickle at 6am, becomes a steadier flow at 7am, rising to a total glut at 8am, before diminishing again to almost nothing by 10am. So long as the number of cars entering City Road is under 10 cars per minute (the 'arrival rate') and the cars are evenly spaced, the traffic lights can cope. The number entering the road each minute can all pass through a single phase of green lights. But while the system can cope with 10 evenly spaced cars every minute, it takes only an 11th car to begin to clog up the system. There begins to be a permanent and growing queue waiting at the traffic light.

We start at 8am with no cars in the queue and the lights turning red.

TIME	Cars arriving in next minute	Cars going through lights in one minute	Queue length when lights go red one minute later
8:00	11	10	1
8:01	11	10	2
8:02	11	10	3
8:03	11	10	4
:	:	:	:
8:20	11	10	21

So in 20 minutes, the queue has built up to 20 cars. In fact the situation will be worse than this. First of all, as rush hour builds up the arrival rate gets higher and higher, so that by 8.20 it could be up to 20 cars per minute with only 10 getting through the lights. Then there s the problem that as the queue length gets longer, it might begin to reach back to the vious set of lights on the road, which means that cars may not even be able to get gh the previous lights when they are green. Add to this the real situation that cars don't arrive evenly spaced but in bunches, and you can see that traffic chaos is on the way.

If the queue of cars at the lights is 25 cars when Brian arrives at the end of it, then instead of passing straight through the lights, he will have to wait for the duration of two traffic light changes before his batch of 10 cars is able to get through. And if traffic light changes happen only every minute, this means he has lost at least two minutes in his journey time. So to go back to the original situation, the reason why John's journey takes twenty minutes less than Brian's is down to the service rates of the traffic lights not being high enough to cope with the extra traffic.

Queues without traffic lights

Queues cause enormous headaches for traffic planners. It isn't just traffic lights that cause them. Any constraint to the free flow of cars could lead to a queue – a roundabout, an accident or roadworks for example. One less obvious cause of a queue is a car slowing down or stopping temporarily in traffic before it drives off again.

You may have experienced this yourself. You are driving at 70mph down the motorway when suddenly the traffic ahead slows down and you pull to a halt, cursing at the prospect of some roadworks or an accident ahead. But after about five minutes of stop start, the cars

One man's fish is another man's Poisson ...

Why is it that at the fishmonger's shop there is sometimes a queue even on a quiet morning? It's all down to what is called the Poisson distribution. If the average number of customers arriving in the shop is one per minute, this doesn't mean that for every minute exactly one customer will arrive. One minute there might be none, the next three, and the next only one. If the arrivals are truly random and there are on average A customers per minute, then the chance of N customers arriving in any given minute is given by a formula:

$$\frac{e^{-A} A^{N}}{N!}$$

N! is N factorial, and this is another guest appearance for the enigmatic number 'e' (see page 130). If there is an average of one customer per minute (A is 1), the using this formula the chance of four people arriving at the fishmonger's shop in any particular minute is about 0.02, or a 1 in 50 chance.
Poisson applies to traffic jams as well as shopping queues, and adds yet more headaches to the traffic planner's life.

in front pull away and suddenly you are back to 70mph. There is no sign of any accident or blockage. It's almost as if there was a phantom crash.

In fact what has happened is that the motorway has reached saturation point. The sheer quantity of cars has messed up the spacing that we all like to have between ourselves and the cars in front in order to feel safe. If for some reason the car in front slows down and you are close behind it, you slow down too. When the car in front speeds up again, it takes a moment for you to react to this, so for an instant the space between the cars gets wider. However, the car behind you is still travelling at your slower speed. And if the car behind *him* was quite close, he too has had to brake.

Picture in your mind a 'shock wave' of decelerating cars, passing along the motorway. If you like, imagine you are holding one end of a big spring – if you shake the spring, you will see the pulse of compressed coils travel along the length of the spring. This is what is happening to the cars. The speed at which this pulse travels can affect whether the traffic grinds to a halt or sorts itself out. To understand this, it is easier to think about another daily queueing problem, the escalator.

Pulses and escalators

ot content with allowing the staircase to do all the work, fast-paced London commuters ve to walk up the moving escalators. It helps to shave valuable seconds off their journey es. But it only takes one tired tourist to stand on the wrong side for the whole marching y to a come to a standstill.

hat happens is this. The person walking immediately behind the tourists suddenly And just like the motorway traffic, a pulse of stopped traffic now shoots down the tor. If the escalator is packed, the stopping might be almost instantaneous all the way the bottom.

suppose that the blockage is cleared and the person at the front of the stopped file s walking up again, like the slowed car on the motorway speeding up.

ok at a diagram and invent some numbers:

The escalator is moving up at two steps per second. At the instant where we have frozen the action, Jo at the front of the blockage, is 10 steps from the top of the escalator. Christine, five places behind her, is 15 steps from the top. Five places behind Christine is Stephen, who is 20 steps from the top. Let's suppose it takes one second for someone to notice that the person in front has started moving and to then step into action. Jo now starts to walk up the escalator.

Now let's wind the clock on five seconds. Five seconds later, the 'pulse' of people starting to walk has moved down five people (one person per second). This means that it has reached Christine, who starts to walk up the escalator again. Since five seconds have passed, the escalator which is climbing at two steps per second has moved up ten steps, so Christine is now just five steps from the top. Stephen, five steps behind her, is still stationary, but the escalator has now carried him to ten steps from the top.

Five seconds later....

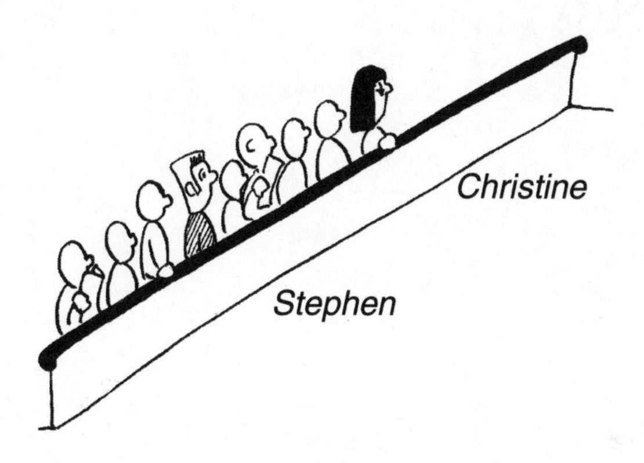

After ten seconds, the moving file reaches Stephen, who begins to step upwards ... but he discovers he is now at the top anyway. The pulse of walking passengers has moved up the escalator to the top – and it now promptly disappears! The rest of the customers remain standing still, and this will continue until there is a gap in the passengers arriving at the bottom of the escalator. And this is all because the pulse of movement was going up the escalator. If the escalator was travelling more slowly, and the passengers were able to react faster to the queue ahead of them beginning to move, then the pulse would move down the escalator and in no time the static queue on the escalator would be full of walking people again.

What this shows is that in any flow of traffic, be that cars or people, there will be pulses of slowing down or speeding up traffic. This pulse will move towards the front of the congestion or towards the back, depending on the relative speeds of the overall traffic and the individuals' reaction times, and this can make the difference between a stoppage which clears itself and one which creates a ten mile jam.

On a busy motorway like the M25, there could be hundreds of pulses passing along the motorway at any one time. Each driver will react to a pulse differently, but a very cautious driver may actually over-react and slow down a lot, and the car behind him may be too close and be forced to stop completely. We now have the start of a traffic jam. A stopped car on the motorway is like a mini traffic light. The arrival rate is the number of cars per second streaming in behind the stopped car. The service rate is the number of cars that can get from a standing start to cruising speed per second – which is invariably slower than the arrival rate (especially on a cold morning when one in ten cars is going to stall in the panic). And this is why, after a seemingly innocuous bit of braking by a tourist who has spotted his turn-off at Junction 21 just a little bit late, the whole motorway can grind to a halt.

Going slower to go faster

How do you make cars go faster on the M25? The answer is to make them go slower.
In 1994, traffic planners decided to experiment with speed limits during rush hour. When traffic on the M25 was heavy, the speed limit was reduced from 70 mph to 50 mph. The result of this was that there were fewer extreme pulses of traffic, and a much smoother flow. Best of all, the number of complete stop/starts was reduced, and this reduced the number of traffic jams. It turns out that the number of cars that are able to feed through the M25 system when it is busy is higher with the speed limit at 50mph than at 70mph.

Shopping queues

Almost as frustrating as a motorway queue is a supermarket queue. The mathematics of supermarkets has a lot in common with that of the roads. If you turn up at Sainsbury's at 4.30pm on Friday evening, you might be able to do your complete shop in twenty minutes. But arrive at 5.30pm, and suddenly your shopping takes you an hour. And the reason? Partly, of course, it is because you are having to play dodgems with your trolley as you desperately try to swerve past the toddlers to reach the tinned tomatoes, but mainly it is because you spend a lot longer waiting in the checkout queues. There are more customers in the shop, so the arrival rate is higher, but the service rate of the checkout assistants may still be the same.

A supermarket does of course have an advantage over the local traffic planners, because if the number of customers goes up, the supermarket can open more tills – which is the equivalent of opening up another street with its own set of traffic lights and increasing its service rate. Supermarkets can also overcome some of the frustration by opening up 'express checkouts' for customers with only a basket.

Interestingly, although it seems fairer to everyone to have one normal checkout and one express checkout, this set-up actually makes the average time spent in queues longer than if all checkouts were the same. The reason for this is that there will be occasions when there are no express customers and so the checkout is lying idle. Because the choice of tills is now restricted for many of the customers, the overall usage of the tills is less efficient. But it's the family shoppers who lose out.

Quirky Queueing Facts

- *On a typical November day, 518,000 cars stop in a traffic jam on the M25 motorway. 29 man years are spent waiting in M25 jams on such a day. The longest queues on this motorway have been over 20 miles.*
- *Queues used to be such an important part of life in Russia that if a Russian saw a queue he would immediately join it and only then ask what it was for.*
- *The simplest formula in queueing theory is for the number of cars in a queue after T minutes. This is given by $N = (A - S)T$, where A is the arrivals per minute and S is the number leaving the queue per minute (S is known as the service rate).*
- *Queueing (when spelt this way) is the only English word with five consecutive vowels.*

The impossible overtaking...

At the beginning of the chapter, we considered the situation of Brian discovering that if he leaves for work at 7.30 his journey time is 30 minutes, but if he leaves home ten minutes later at 7.40, his journey time is 50 minutes. Brian has been thinking about this. He has discovered that if he leaves home not ten minutes later but an hour later (at 8.30), his journey time is only 20 minutes. So it is possible that the later you leave, the less time it takes to reach the destination. Does this mean, wonders Brian, that it might be possible to find a time in the morning when leaving a minute later could lead to him cutting the journey time so much that he actually arrives a minute earlier? This is the lazy commuter's dream! Perhaps you can explain to him the fault in his logic.

17

WHY ARE SHOWERS ALWAYS TOO HOT OR TOO COLD?

From squealing microphones to population explosions.

You are on an overnight trip to a big city and you decide to stay in a cheap hotel. You know it is a mistake when you discover that the TV in the bedroom needs to be slapped twice to get a stable picture, and you open the curtains to find that your room overlooks the bus station. The lowest point of all, however, is when you climb into the shower, turn it on and then leap straight out as a jet of freezing water hits you in the face.

Scrabbling to recover the situation you turn the tap round to HOT, test the water and find it is only lukewarm, and so turn up the temperature to MAXIMUM. At last you sigh with relief as the water hits a pleasant temperature, but as you begin to soap up, the temperature soars beyond the comfort zone to scalding point. You slam the control down but it's still too hot, so you force it down to MINIMUM. Soon the water is cold again, and you reverse the process. You have entered a cycle of hot/cold which is only broken when you finally climb out of the shower cursing the day you booked into the place.

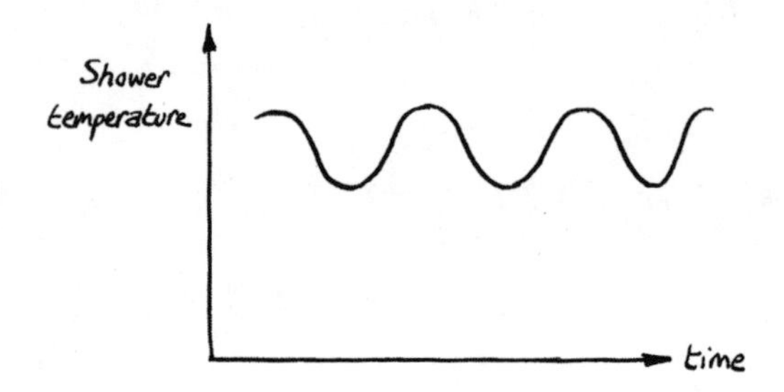

Why is it that hotel showers are never at the right temperature? The explanation is one that links showers with plagues of insects, economic recessions, automatic pilots and squealing microphones. It is all down to *feedback*, or the fundamental interconnectedness of all things, as Douglas Adams' holistic detective Dirk Gently would have put it.

Actions and consequences

If somebody gets angry and throws a punch, they must be very naive if they don't expect to get a punch in return. It was the mathematician Newton who first recognised that every action has a reaction, although he wasn't talking about pub fights at the time. Newton was referring to physical forces, but the principle of action and reaction applies to almost any situation. If the reaction has some effect on the person who initiated the action, this is known as feedback. The input (first punch) leads to a response (feedback is a return punch) which has a direct effect on the inputter (he decides to throw an even bigger punch).

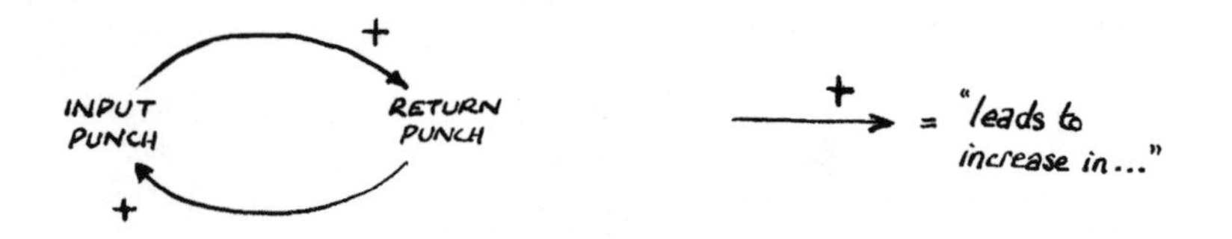

In fact, this is an example of *positive* feedback, although it might not feel very positive at the time. Positive feedback is when the response to an action causes the initial action to increase. If you've ever heard a microphone at a rock concert suddenly let out an ear-piercing squeal then you have heard some positive feedback which was caused by the microphone (which is the input) being too close to the speakers (the output). The speakers feed back to the microphone.

Positive feedback is also seen in population growth.

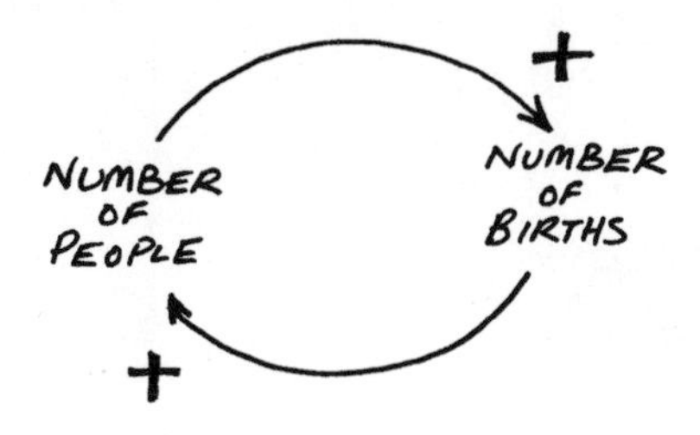

A larger population means more children which means a larger population which means more children. All of this may lead to . . . .

Exponential growth

If a population is small, the rate of growth of the population is often proportional to its size. This means that growth will be *exponential.* In fact more than that, the graph of population size will belong to a very particular formula, which is $P = P_0e^{Bt}$, where P is the population size, P_0 is the starting population, B is the birth rate and t is the time. The number 'e' is just under 2.72 and is one of those special numbers like π and ϕ which crops up all over the place (see the box over the page).

Exponential growth is what Australia experienced in its population of rabbits. Rabbits were introduced into Australia because farmers wanted to have something to shoot for fun. Unfortunately rabbits found Australian conditions so much to their liking that within a few years they had become a national pest.

Exponential lilies

In a village pond, the lilies grew so fast that they doubled the area that they were covering every day. After 30 days the entire pond was covered. After how many days was the pond only half covered?

The answer to this old riddle is of course 29 days. That's the dramatic effect of exponential growth. In this case the area covered by the pond was a formula along the lines of $A = 2^t$, where t is the number of days and A is the area covered.

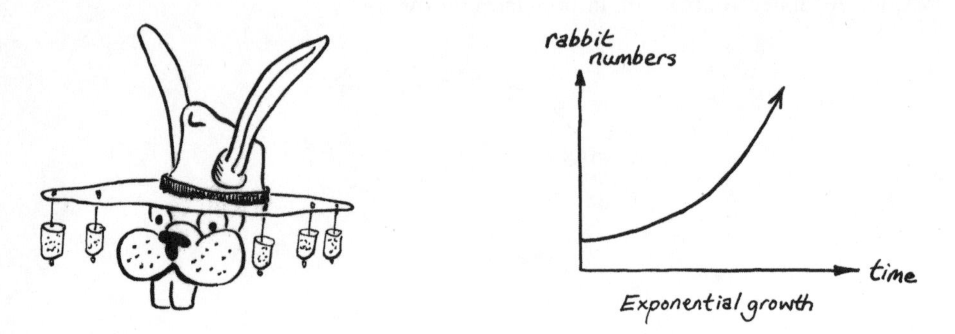

If exponential growth was allowed to go unchecked, then the world would be overrun in no time. Fortunately, there are other factors which limit population growth. These factors are *negative* feedback.

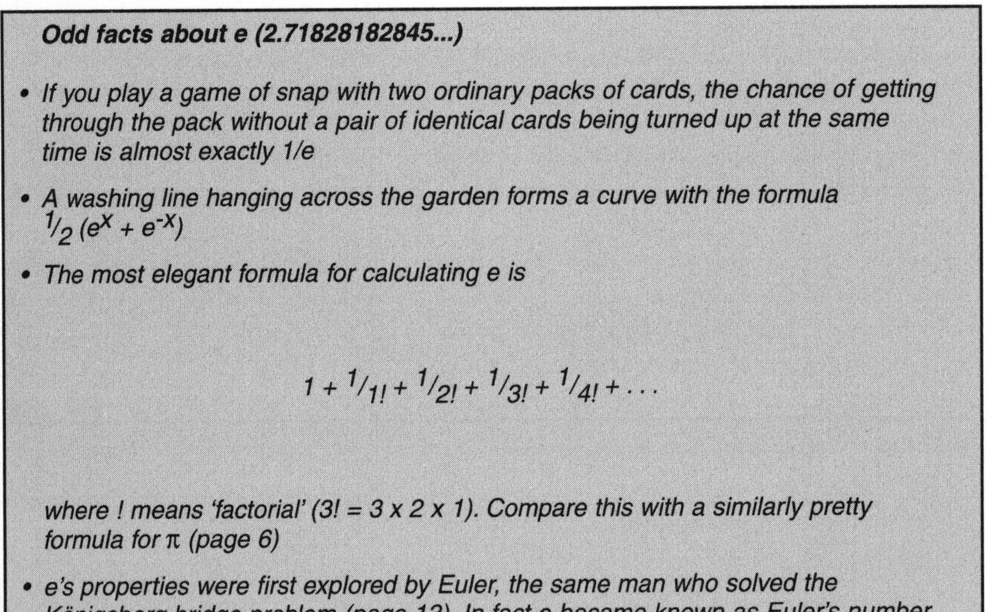

Odd facts about e (2.71828182845...)

- *If you play a game of snap with two ordinary packs of cards, the chance of getting through the pack without a pair of identical cards being turned up at the same time is almost exactly 1/e*
- *A washing line hanging across the garden forms a curve with the formula* $\frac{1}{2}(e^x + e^{-x})$
- *The most elegant formula for calculating e is*

$$1 + \frac{1}{1!} + \frac{1}{2!} + \frac{1}{3!} + \frac{1}{4!} + \ldots$$

where ! means 'factorial' (3! = 3 x 2 x 1). Compare this with a similarly pretty formula for π (page 6)

- *e's properties were first explored by Euler, the same man who solved the Königsberg bridge problem (page 13). In fact e became known as Euler's number, though it is a coincidence that it was Euler's initial.*

Negative feedback and control

As a car driver approaches a sharp right hand bend, his instinct tells him to jerk the steering wheel to the right. The turn of the steering wheel is his input. His eyes now provide him with feedback. If he hasn't turned the wheel far enough, then his brain sends a message to his hands to move the wheel further. But if he has turned the wheel too far, the brain sends a message to make a correction in the reverse direction. A healthy, alert driver will be extremely responsive, and very swiftly the steering wheel will be in exactly the right position. The adjustment to the angle of the steering wheel has probably looked like this:

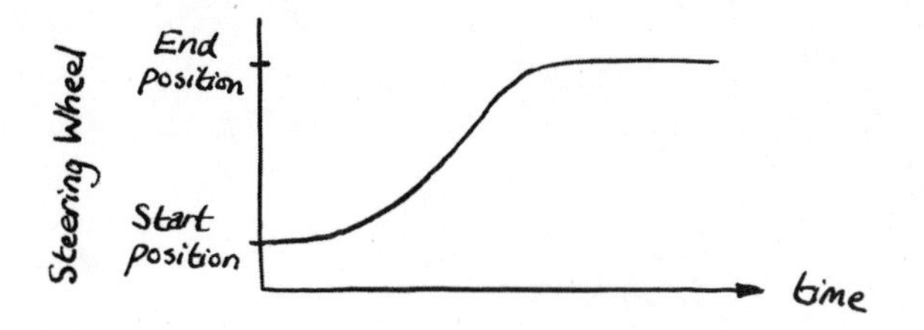

Notice how it shifts rapidly to the correct level. A learner driver might slightly overshoot leading to a small fluctuation around the correct position which rapidly damps away to nothing:

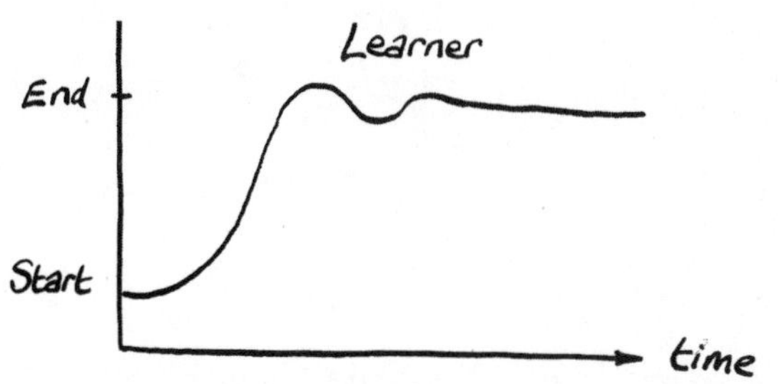

This was an example of a system with negative feedback built in, which allowed the driver to move from one fixed position of the wheel, known as a steady state, to a different one. The negative feedback applies when the wheel has turned too far. However, the consequences are not always so calm if the negative feedback mechanism is wrongly adjusted. This can happen if there is thick fog.

In this case, not only might there be a slight delay before the driver realises that the corner is there, but his reactions might also be more extreme. He is more likely to overshoot when he turns the wheel, and more likely to over-react once he realises he has overshot. This is an example of a system which is over-responsive.

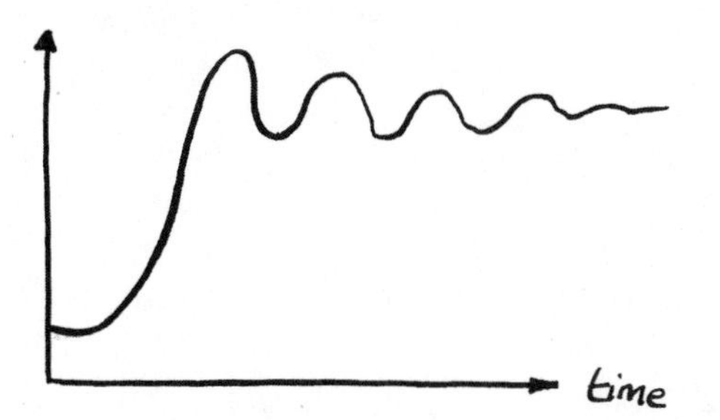

In this case there is a happy ending because the driver gets the right amount of turn eventually. However, the wrong level of reaction can lead to the driver losing control of the car.

Foxes and rabbits

Let's return to the wild, where graphs like those for the car driver can be seen in the size of animal populations. The Australian rabbit population grew exponentially because there were no rabbit predators in that country. The rabbit numbers would have been very different if there had been enough foxes to keep their numbers down.

Predators are one of the main factors that limit an animal's numbers. Another factor is food. Usually the greater the population of animals, the less food there will be to share per head of population. The less food there is, the greater the hunger and hence the higher the death rate. This is what is called a *negative feedback loop.*

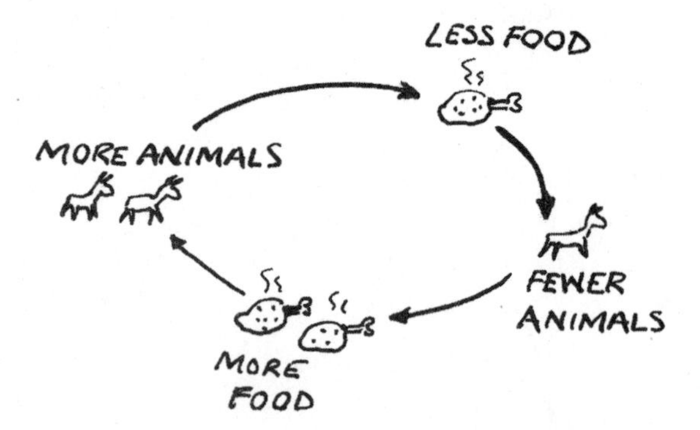

What happens to the predators, though? Their food, or prey, is other animals. In this situation, the predators want to eat as much prey as possible in order to survive, but if they are too successful they will end up doing themselves out of their only source of food. Somehow there has to be a control which stops the prey from going too far.

To see what happens to predator and prey populations, it is easiest to create an artificial world in which only two types of animal exist, foxes and rabbits. In this world, rabbits are the only food that foxes eat.

One way to model this world is to create a formula for the birth rate and death rate of foxes and rabbits (that is, the number of animals born and dying each month). A highly simplified pair of formulae for the number of foxes F and rabbits R each month might be:

$$\text{New } F = \text{Old } F \times B_f - F \times D_f / R$$
$$\text{New } R = \text{Old } R \times B_r - F \times D_r \times R$$

Old F is the total number of foxes, and Old R the number of rabbits last month. New F and New R are the numbers of foxes and rabbits this month. B_f and B_r are the monthly birth rates of foxes and rabbits (measured in babies per animal per month), while D_f and D_r are the natural death rates for foxes and rabbits.

The formulae are based on common sense. In the case of foxes, the number of deaths goes up if the number of foxes increases or the number of rabbits goes down, since both of these mean that there is less food available per fox. For rabbits, the number of deaths goes up as the number of foxes goes up.

What these formulae lead to is a cycle. At the starting point, foxes might consume a lot of rabbits, and so fox numbers go up while rabbit numbers go down. However, after a period of glut the foxes will begin to die of starvation due to competition for a limited food supply, and so both populations will be in decline. When fox numbers hit a low, rabbit numbers recover, and before too long fox numbers are able to increase as well. Although the exact pattern depends on how quickly foxes and rabbits breed and die off, the numbers in the two populations might fluctuate like this:

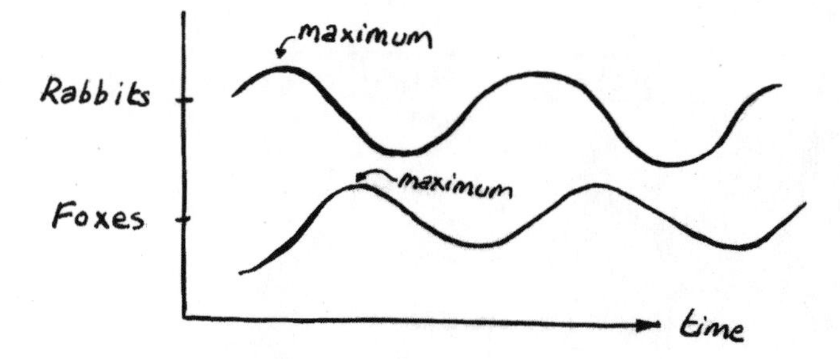

Another way to see what is happening to the number of foxes and rabbits is to plot these two numbers against each other over time. If the populations fluctuate steadily, as in the previous graphs, then the numbers of foxes and rabbits will move in a cycle like this:

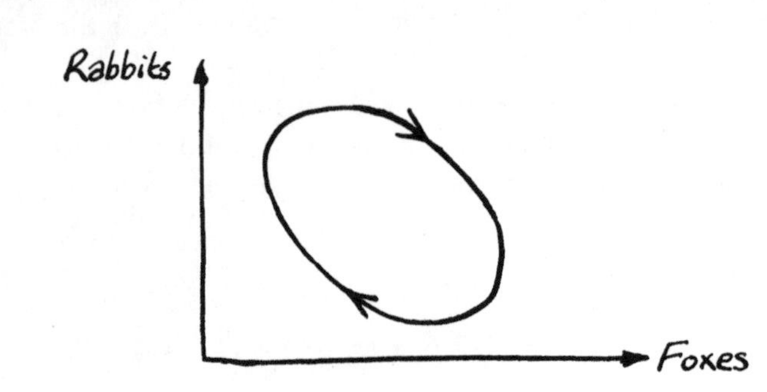

There will be times when rabbits appear to be in permanent decline, but this decline suddenly halts as foxes themselves go into decline.

Nature appears to be quite resilient to big changes which could throw this cycle out. For example, an extremely cold winter might kill off far more foxes than usual, but the result of this is that those foxes that remain will have an even larger supply of rabbits to feed from the next year and hence their growth rate will be stronger. In most situations a steady state is restored. If it isn't, then there will be one of two long term effects: a catastrophic explosion in numbers because of a shortage of predators, as happened with the Australian rabbits, or a permanent decline in numbers which leads to extinction. Two factors seem to have the biggest influence on throwing the cycles out in this way: environmental disasters such as the meteorite believed to have wiped out dinosaurs, and human intervention.

Fluctuations of fox and rabbit numbers

Below is a table showing how fox and rabbit numbers fluctuate if the numbers under the table are put into our formula. The story starts with 100 foxes and 100 rabbits. Notice how rabbits suffer a catastrophic decline after month 2, and fox numbers plummet after month 8, when rabbit numbers have already begun to rise again.

Month No.	Fox number	Rabbit numbers
0	100	100
1	110	90
2	120	78
8	100	23
16	14	39
24	18	126
32	54	290
40	178	125

The values used in the formula for birth and death rates were:
$B_f = 1.2$, $D_f = 10$, $B_r = 1.2$ and $D_r = 0.003$

The 1988 slump: too much 'hot tap'?

In 1989 Britain's booming economy took a sudden turn for the worse. One explanation for this was that the economy was over-heating in 1988. All of the indicators (such as employment and inflation) were looking a little on the cool side, so the Chancellor believed that the economy could do with more of a boost by dropping interest rates and taxes even further. What he didn't realise was that, like the hotel shower, the boom in the economy (hot water) was already on its way, and he was stoking this up even further. The result was that the economic temperature shot through normal to very high. A huge balance of payments deficit appeared as Britain sucked in imports, and the Chancellor had little choice but to slam on the cold tap by putting a huge increase in interest rates. The sharp downturn in the economy followed, and all because the Government had assumed that the economic hot water tank was more responsive than it turned out to be.

Time lags and the shower

So what is the explanation for the hot and cold shower problem described earlier? The shower has a feedback system. The input is the level to which you turn the hot/cold tap. Your skin then senses the output, which is the water temperature. If the temperature is too low, then you adjust the tap accordingly. This is the feedback.

The problem is almost certainly that the shower operator wrongly interprets the feedback that he is getting. He assumes that the temperature of the water that he is feeling is immediately related to the amount by which he has turned the tap. However, the truth is that the hot water tank may be some distance away, which means that there is a time lag between turning up the temperature and feeling the result.

This time lag is similar to the time lag in the deaths of foxes. Foxes don't die immediately when the rabbit population disappears, but a month or so later. If the shower operator isn't careful, he can find himself in the same cycle as the foxes and rabbits.

One of the most worrying examples of an over-reactive system has been the media. There was a time not long ago when a news story took several days to develop. There was less demand for immediate analysis, and therefore the reporters were able to digest a story before giving their opinions. Now an analysis of a story and the reaction from all the parties involved is almost instantaneous. This can lead to stories which appear to oscillate from one extreme to another. Reports on disasters typically start with an underestimate ('at least fifty people have died ...') climb to an over-estimate ('as many as 400 are now feared dead ...') before reaching the final stable answer somewhere in between ('it is now known that 241 people were killed ...').

Society has created, and is now the victim of, a system which is out of control. Any wise shower operator knows the solution to this. You get to the right answer sooner if there is a pause for reflection before turning the tap.

18

HOW CAN I GET THE MEAL READY ON TIME?

Critical paths and other scheduling problems.

LET'S TALK ABOUT FOOD AND FUEL-SAVING COOKERY!

That was the headline of an article in *Popular Women's Weekly* in November 1941. It was part of a wartime campaign to avoid wasting anything, particularly in the kitchen.

Even toast played its part in the wartime effort.

Toast? An advertisement appeared in another magazine advising housewives on how to make three pieces of toast more efficiently. Mrs Smithers had a gas grill which would take two pieces of toast, toasting one side at a time.

Mrs Smithers wanted to make three pieces of toast, one for Dad, one for herself, and one for Smithers Junior. The obvious way to make three pieces of toast (which we will call A, B and C for convenience) is as follows:

Put pieces A and B under grill and toast top	(30 seconds)
Turn over and toast other side	(30 seconds)
Remove A and B and put in C	(30 seconds)
Turn over C	(30 seconds)

Overall this meant two minutes of grilling time. But wait! Researchers saw that there was some inefficiency in this approach, and with a bit of reorganising, they could shave 25 per cent off the energy required to make the Smithers' toast:

Put pieces A and B under grill and toast top	(30 seconds)
Turn over A and replace B with C	(30 seconds)
Replace toasted A with B and turn over C	(30 seconds)

Three fully toasted pieces of toast in just ninety seconds.

Domestic Britain was being educated in the art of scheduling and *critical path analysis*, or 'how to get your project done in the minimum time possible'. In fact it wasn't called critical path analysis then – that name did not appear until the 1950s – but anyone who had ever had to prepare a dinner with several courses already knew that there was a right and a wrong order in which to do all of the operations if they were to be ready by tea time.

Getting the order right

There are some tasks where there is only one possible order of doing things. Every child knows that socks go on before shoes, for example. Shoes and socks are *sequential* activities, shoes being dependent on socks.

However, there are other routine tasks where there is a choice in which order to carry out tasks. For example when having a bath, there is the choice of:

A Get undressed and then turn on the taps
B Turn on the taps and then get undressed

No disaster awaits if you choose A. It does, however, mean that you will end up getting into

The bridge problem

Four men need to get across a foot bridge to catch the last train which leaves in under 16 minutes time. But there is a catch. The bridge can only hold two people. Because it is hazardous, the men crossing need a torch at all times, and to cross together at the speed of the slower one.

James can cross the bridge in 1 minute.
Keith can cross it in 2 minutes.
Larry can cross it in 5 minutes.
Mick is very nervous, and takes 8 minutes to cross it.

How can all four men get across the bridge in time to catch the train? The men have just one torch between them, and the torch can only be carried by hand, it can't be thrown. If Mick crosses the bridge with Keith, say, and Mick then brings the torch back to the others, that takes 16 minutes, which has already missed the deadline.

The answer may seem counter-intuitive:
James & Keith cross first (2 minutes)
Keith brings torch back (2 minutes)
Larry & Mick cross together (8 minutes)
James brings torch back (1 minute)
James & Keith cross together (2 minutes)
They get across in 15 minutes.

the bath slightly later. If it takes two minutes to undress and ten minutes to run the bath, then option A takes 12 minutes. Option B will only be 10 minutes. This is because undressing and running the bath do not depend on each other and can be carried out in *parallel*.

Sorting out the sequential tasks (*e.g. putting water in the kettle before switching the kettle on*) and the tasks which can run in parallel (*listen to radio news while doing the ironing*) is at the heart of running a household. On a grander scale, sequential and parallel tasks are the important parts of critical path analysis which is used by project managers throughout industry.

Shepherd's pie and the critical path

Here is an example of a project. Shepherd's pie is the meal a bachelor graduates to when he has mastered the art of beans on toast. Steve has decided to cook shepherd's pie tonight, but seeing that there is a lot to get done, he asks flatmate Craig to help out. The televised football match begins in 40 minutes, by which time they want to have the food ready. They have two gas burners and an oven, plus a large, deep frying pan, a saucepan and a casserole dish.

Here are the tasks facing the two lads in the kitchen and the time each one will take. (Some of these times are debatable, but let's assume this is Steve's recipe and timetable and he's sticking to it).

A	Prepare spuds (wash, peel etc)	7 minutes
B	Boil water	3
C	Boil spuds in water	17
D	Mash spuds	3
E	Chop onions (and bathe eyes)	4
F	Fry onions	3
G	Brown mincemeat	5
H	Add boiling water to gravy cubes and pour into mince	2
I	Simmer mince and put in casserole	11
J	Smooth mashed spud on top of mince	2
K	Heat oven	5
L	Put pie into oven and grill	8

There are two main operations involved here, one for the mince and the other for the spuds. These two can go on in parallel.

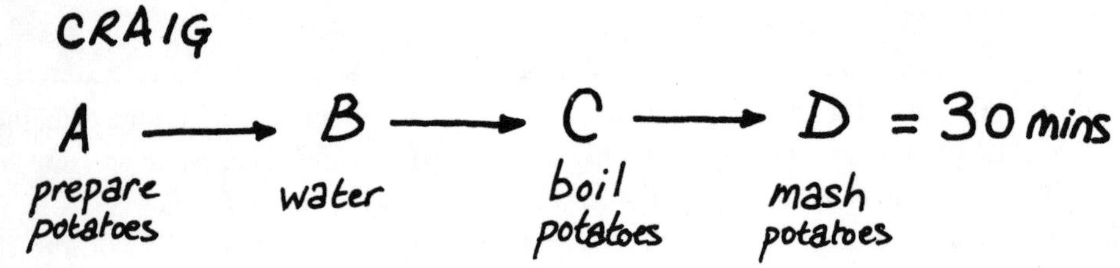

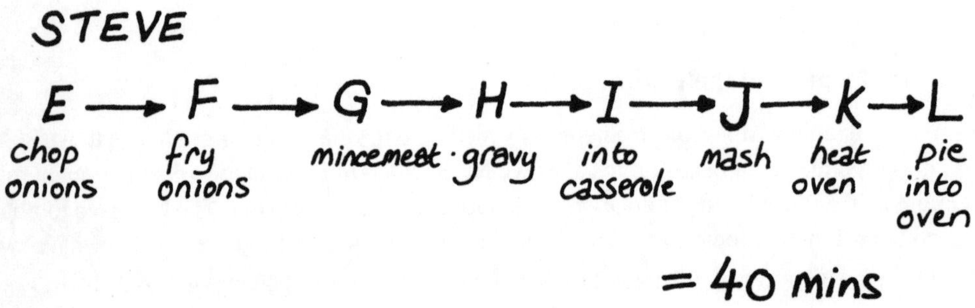

CRAIG

A → B → C → D = 30 mins

A: prepare potatoes; B: water; C: boil potatoes; D: mash potatoes

STEVE

E → F → G → H → I → J → K → L

E: chop onions; F: fry onions; G: mincemeat; H: gravy; I: into casserole; J: mash; K: heat oven; L: pie into oven

= 40 mins

At first glance it looks as if the meal will be ready within the 40 minute deadline. Unfortunately there is a problem. Task J (mash onto mince) can only happen when the mash is ready (D). The mash is not ready until 30 minutes into the cooking, even though Steve could have reached point J after only 25 minutes. So in fact the pie cannot be ready for at least 45 minutes. The match will have started by then!

This is a classic case for some critical path analysis. Just how quickly can Steve and Craig cook the shepherd's pie?

There is a technique for finding this out which is shown in the diagram below. The trick is to set out the tasks which are sequential and those which can happen in parallel. In this case there are five tasks which are not dependent on any others. These are A (prepare spuds), B (boil water), E (chop onions), G (brown mince) and K (heat oven). These can go down the left hand side, and to the right of them come the tasks which are dependent on previous ones.

There are various interconnections here, such as H (gravy mix) having to come after the water has boiled (B). Browning the mince and frying the onions can happen in either order, but in Steve's recipe they can't happen at the same time. It makes sense for the mince to be browned first, since this means that the onions can be chopped while the mince is browning.

Following the whole process through to the finished pie, it turns out that the meal can

Critical path of a shepherd's pie

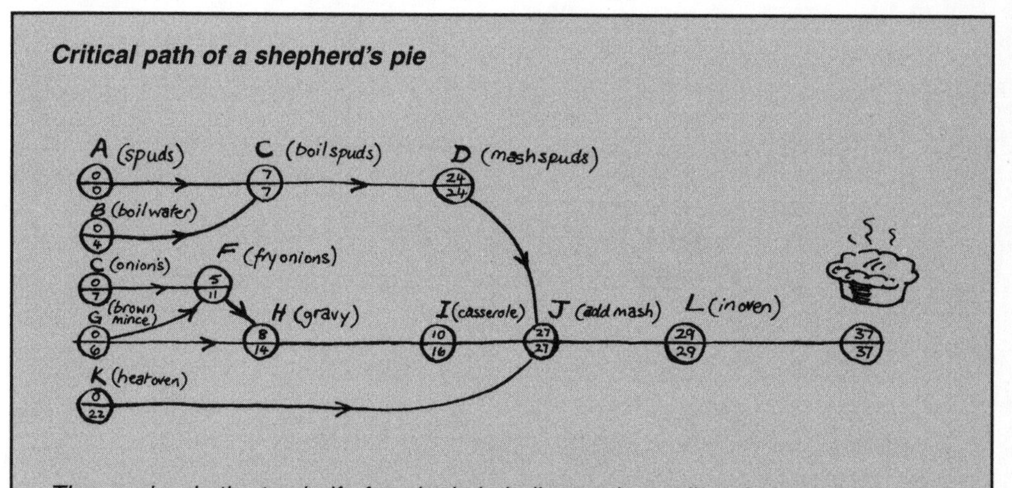

The number in the top half of each circle indicates the earliest time at which a task can begin, and the number in the bottom half indicates the latest time that it can begin. The earliest time that a task can begin is found by working through the tasks from left to right. The latest times can then be worked out by starting from the finished pie and working backwards.
For example, boiling the water (B) can start after zero minutes, since there are no tasks which need to come before it. The latest that it can start if it is not to delay the whole meal is four minutes (the lower number in its circle).

be cooked in a minimum of 37 minutes. If this 37 minute deadline is to be met, however, there is one sequence of tasks in which earliest and latest possible start times are the same (see the diagram). This sequence is: Prepare spuds, boil spuds, mash spuds, add spuds to mince, place the concocted pie into the oven.

This is the critical path. Any delay to any task in this series will delay the shepherd's pie. On the other hand, any task which can be speeded up in this critical path will shorten the cooking time. Steve can prepare the spuds in less than seven minutes if he lops off a bit more of the flesh than normal or leaves some of the skin on the potato. Four minutes trimmed off this operation would bring the entire meal cooking time down to 33 minutes.

There is one slight complication which the critical path diagram does not take into account. At one point there are five tasks going on at once: A B E G and K. This is not possible with just two cooks. However, frying the onions and browning the mince are not full time activities, and they do also belong to the 'slack' side of the meal (not on the critical path) so in practice the lads should just about get away with it. They can watch the football thanks to the wonders of critical path analysis.

Reduced waiting times

Sometimes shuffling the order of tasks makes no difference to how long it will take to complete them all. This doesn't mean that the order is irrelevant, however.

One day a surgeon has to operate on five patients. The patients need different types of operation requiring different amounts of surgery time. These are:

Patient	Time of Operation
Adam	30 minutes
Barbara	120 "
Claire	90 "
David	80 "
Ernie	75 "

No matter what order the surgeon carries out the operations the total elapsed time will be the same, so there is no prospect of him getting a quick game of golf in the afternoon. However, the order *will* affect the *average* waiting time for each of the patients, so he can still influence his customers' satisfaction.

Suppose that the surgeon decides to conduct the operations in the sequence A, B, C, D, E. Since A is in the operating theatre for 30 minutes, B must wait 30 minutes before her turn. After A and B, 150 minutes have elapsed, which is C's waiting time. D will have to wait a total of 30 + 120 + 90 minutes, while E will have to wait a total of 30 + 120 + 90 + 80 minutes.

The waiting times of A, B, C, D and E are:

Patient	Waiting Time
Adam	0 minutes
Barbara	30 "
Claire	150 "
David	240 "
Ernie	320 "

The average waiting time of each patient works out as 740/5 or 148 minutes.

Now see what happens if the order of patients is changed so that the one with the shortest surgery time is first, the one with the next longest time is next and so on. The sequence in which the surgeon operates is now A, E, D, C, B.

Patient	Surgery Time	Waiting Time
Adam	30 minutes	0 minutes
Ernie	75 "	30 "
David	80 "	105 "
Claire	90 "	185 "
Barbara	120 "	275 "

On average each patient is now waiting only 119 minutes instead of 148 minutes. There has been an increase in patient satisfaction and all the surgeon has had to do is schedule his patients in a better order.

Big projects

Steve's shepherd's pie and the surgeon's operation shuffle are small scale versions of many major projects. Construction projects, manufacturing processes and military operations – indeed any project which involves many activities taking place at the same time, will these days rely on a project manager using critical path analysis. He will almost certainly be aided by a computer.

There are many complications that can add to the simple cases that we have looked at which make computer power essential. For example, what if Steve and Craig had at one point in the cooking needed three gas hobs? The tasks would have to be shuffled to accommodate this. Or suppose that Steve's mum has a habit of ringing up at about this time every evening. If she did tonight, it would slow down Steve's part of the cooking. Steve's mum is a risk that really ought to be programmed into the planning of the project. On a construction site, the equivalent of Steve's mum can be the risk of bad weather, for example.

A sophisticated project planner can build risks into critical path analysis and probably trim 25 per cent off the project cost by scheduling the tasks more efficiently. Those involved in theater or movie production could take note of this. After all, think of all those last-minute crises that hit the producers when the sensitive lead disappears from the set after a tantrum, or when the rain sets in just before a summer scene. A project planning program could reduce these crises to a minimum. Then again, maybe that would take away some of the adrenaline. There's nothing that the dramatic professions seem to thrive on more than a bit of real-life drama—even if it is self-inflicted.

19

HOW CAN I ENTERTAIN THE KIDS?

Numbers can be magic.

One of the best ways of introducing children to a subject is to attract them with a bit of fun. Nothing achieves this better than magic tricks, and it is hard to find a child under the age of eleven who doesn't like watching a trick. Indeed, most adults are secretly captivated by tricks as well. Math is full of curiosities which can be used as the basis of tricks. Perhaps this explains why so many magicians are also keen on math, and it is no coincidence that Lewis Carroll the great mathematician and children's author also loved magic and puzzles. Perhaps more math teachers should become magicians.

Here is a small selection of tricks. They have all been successful when entertaining children, but they often work with adults too, and indeed the first one was used at the start of a management conference. The conference was a serious one on management skills, but at the end one of the directors said: "I've got just one question ... can you explain how you did that trick at the start of the session?"

Trick 1: Animal magic

You are about to have your mind read. You need to be able to do your nine times table and some simple adding and taking away.

- Think of a number between 1 and 10. Don't tell me.
- Multiply it by nine.....have you got the answer?
- Now your answer is probably a two digit number. Please add up the two digits to give a new answer (for example if it was 25, the digits 2+5 add to 7).
- OK, take away four from this new answer to get your final number.
- Now turn this final number into a letter: 1 is A, 2 is B, 3 is C, 4 is D and so on ...
- Think of an animal beginning with your letter.
- Are you ready? And the animal you are thinking of now is ... .an elephant.

Amazing. But how does it work?

The simple principle is that all the numbers in the nine times table have digits which add to nine. 18, 27, 36, 45 ... they all work. (In fact this works for every multiple of 9 up to 20, except for 11.) This means the assistant is forced to come up with 9, so when he subtracts 4 he gets 5, which becomes the letter E. And how many animals do you know that begin with an E?

Trick 2: The anti-psychic trick

Remove seven cards from a 52 card pack. (It's probably best if they are consecutive and in the same suit, for example the Ace,2,3,4,5,6 and 7 of Hearts.) Ask your assistant to check that these are ordinary cards. Now ask her to shuffle the cards, then take them back and shuffle them yourself. Secretly check the card on the bottom of the pile – let's suppose that it is the Ace of Hearts.

Now tell the assistant that you have strong psychic powers, which enable you to prevent her from picking the Ace of Hearts. Give her the pile (face down), and ask her to think of any number between one and six. Suppose she picks four. Now tell her to count three cards from the top of the pack on to the bottom, one at a time, and then to turn over the top card. Predict that it won't be the Ace of Hearts, and sure enough it isn't. Ask her to place this card *face up* on the bottom of the pack and then repeat the exercise, counting three cards from the top on to the bottom one at a time, and turning over the fourth. She does this routine six times, and each time the card she turns over is not the Ace of Hearts. Only one card is left face down, and you tell her that, as usual, you managed to keep the selected card from appearing until the last moment. Turn it over to reveal the Ace of Hearts.

Prime numbers in Trick 2

The only requirement in this card trick is that the number of cards in the pile is a prime number. In this case it was 7, but the trick would work for 3,5 or 11 cards (any more than that and it starts to get a bit tedious). If the number of cards in the pile is N, you ask the assistant to pick a number between 1 and N-1 (so for 11 cards, the assistant picks a number between 1 and 10).
Suppose the assistant picks 4 and there are 11 cards in the pile. How many times do you have to add 4 before you get to a multiple of 11? Try it. 4, 8, 12, 16, 20, 24, 28, 32, 36, 40, 44. That is eleven turns. What if the assistant picks 6? 6, 12, 18, 24, 30, 36, 42, 48, 54, 60, 66. Eleven again. In fact it will always be eleven. And as long as the number of cards is a prime number P, the number of cycles it takes to get to the bottom card in the pack will always be P – in other words it will always be the last card turned over. To anyone familiar with the principle of prime factors this result may be blindingly obvious, but it makes a surprisingly effective trick, even when performed on mathematicians!

Trick 3: Number forecast

This is another 'psychic' trick which depends on a simple mathematical principle. Prepare four cards on which you have written the following numbers.

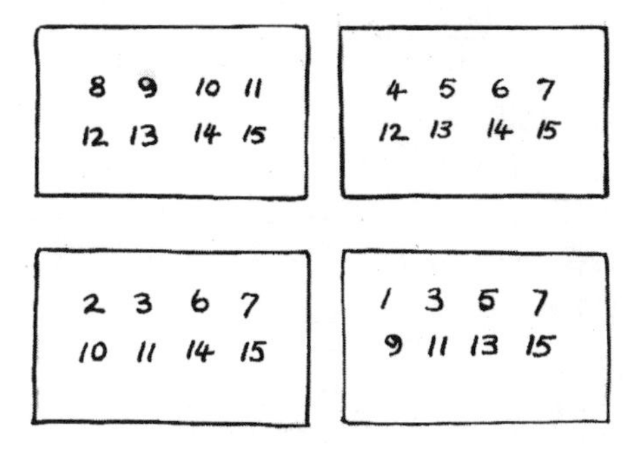

Ask your assistant to pick any number between 1 and 15. Then show him each of the four cards and ask whether his number appears on the card. You then immediately reveal what his number is.

The secret to the trick is simple. Add up the top left numbers of the cards on which the assistant said his number appeared. For example if he chose 13, this appears on the first, second and fourth cards, and you add 8, 4 and 1 to give 13.

Children *love* this trick because they can make their own copy of it very quickly and then try it out on their parents. The trick is also a good introduction to binary numbers (see the box) which are fundamental to computers.

Binary numbers

In trick 3, the number 13 appears on the four cards in the pattern 'Yes, Yes, No, Yes'. Number 8 would be 'No, Yes, No, No'. In binary code, 'Yes' is represented as a 1. 'No' is represented as a 0. The number 13 in binary code is 1101, while 8 is 0100 (in practice you can leave off the initial zero to give 100). Binary numbers work in just the same way as ordinary numbers, except that instead of columns for units, tens, hundreds, thousands etc., the columns are units, twos, fours, eights, sixteens, etc.

Why do computers use binary numbers rather than decimals like the rest of us? The main reason is simplicity. Decimals require us to learn ten different digits, but binary numbers are extremely convenient because they only need two digits. This also means that they are easy to represent electronically. 'On' is 1 and 'Off' is 0.

Trick 4: The Magic square

12	8	5	9
17	13	10	14
11	7	4	8
13	9	6	10

To prepare this trick, make a large copy of this square and have ready four crayons of different colours (let's say red, blue, green, yellow). Next, write the number 39 on a piece of paper and seal it into an envelope. Hand the envelope to a volunteer. You can use four other volunteers for the trick, if you wish. Give the first volunteer the red crayon, and ask him to pick any row and put a red line through it. Now ask him to pick any column and put a red line through that.

The next volunteer gets the blue crayon, and has free choice to knock out any one of the remaining rows and any one column. This is followed by green. The yellow crayon has to go through the only remaining row and column.

Emphasise that there has been complete freedom of choice. Now add up the numbers on the squares where the two red lines meet, the two blue lines meet, the two green lines meet and the two yellow lines meet. This total will be 39, and you can now ask your final volunteer to open the envelope to reveal your prediction!

How was the square constructed? Put these numbers outside the grid. Note that they add up to 39.

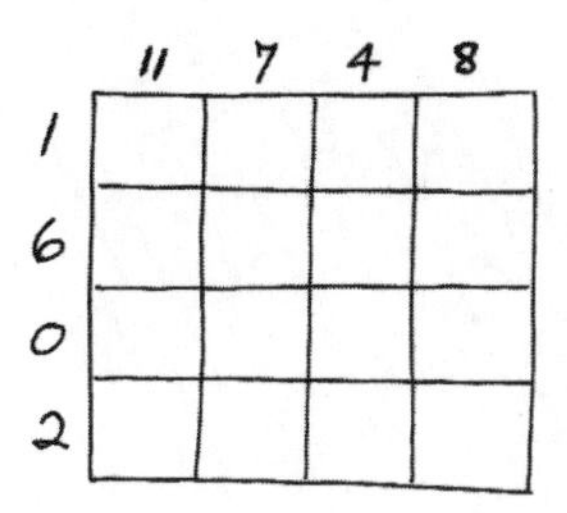

Now fill in the grid by adding the number at the top of its column to the number by its row. This produces the magic square. The crayon crossings out ensure that one number from each row and one from each column is selected, so that the grand total must be the same as the total of the numbers used to create the square.

You can make a square like this to fit any 'magic' number you wish. If you have a relative who is about to have their 50th birthday, you could now produce for them a special birthday square which always produces 50 by making sure the numbers around the grid add up to that number. That's assuming that they want to be reminded of the fact.

Trick 5: Boring ingredients, nice result!

This trick requires a calculator.

Prepare five pieces of card with what you describe as a 'boring' number on each one. The boring numbers are 3, 7, 11, 13 and 37. You explain that often in life you can have lots of boring things to do, but it is worth while in the end because the result can be very exciting. Ask a volunteer to shuffle the cards. Now ask her to think of any number between 1 and 9. Ask her to pick one of the cards, and multiply her secret number by the number on the card. Now tell her to pick another card and multiply the result by the number on that card. Repeat this for all of the cards. Before she presses the '=' button, tell her that her secret number is going to suddenly appear before her eyes lots of times. Sure enough, if her number was 5 (say) the result is 555555.

This works because $3 \times 7 \times 11 \times 13 \times 37 = 111, 111$. As it happens, 3, 7,11, 13 and 37 are all prime numbers – they are called the *prime factors* of 111,111. Of course it doesn't matter what order you multiply them together, they always produce that interesting number. If they are multiplied by any number between 1 and 9, the chosen number will appear six times in the answer.

There are two variations on this trick. You could start by using just the 3 and 37 cards. Multiply these by the volunteer's number (say it's 5) and the result is 555.

Next you can use just the 7, 11 and 13 cards, and ask the volunteer to think of a number between 100 and 1000 (say 123). Multiply all the numbers together and you get the assistant's number twice, 123123.

The original trick with all five cards can then be the 'finale'. Kids are usually *very* impressed.

Trick 6: Reversing numbers

This trick also needs a calculator.

- Ask the assistant to think of a number between 100 and 999 (for example 791)
- Now reverse it (197)
- Ask the assistant to find the difference between the first number and its reverse. This is a new number (791 – 197 is 594).
- Find the reverse of the new number (495) and add the two together (495 + 594).

You now produce an envelope in which you have written a number. Get the assistant to say what her final number is, then open the envelope to reveal the answer, which is always 1,089.

Actually, to be absolutely sure of success with this trick you need to ask the assistant to make sure that the first and last digits of her start number differ by at least 2 (so for example 128 is fine, but 192 would not be).

The trick works because any 3 digit number minus its reverse is a multiple of 99. To see why, let's call the number abc. This is the same as 100a + 10b + c. The reverse is 100c + 10b + a. Take the second from the first and you get 99a – 99c, which has to be a multiple of 99. Meanwhile, any multiple of 99, from 198 to 891, when added to its reverse comes to 1,089. Try it and see.

You can add a nice twist to this trick. Start by saying that you are going to send a message to the assistant. Perform the trick as before, but instead of producing the envelope when she reaches the number 1089, ask her to add 200 to her answer, divide the result by 10,000, and then multiply by 6. Tell her that the message is now on the calculator. All she will see is 0.7734. But then say "Oh, I forgot, this is a trick about reversing," so she needs to hold the calculator upside down. Sure enough, there is the word HELLO.

A final word

We saved the chapter on magic until last for a reason. Magic tricks demonstrate one of the most important practical uses of maths, which is to make life more fun. And fun doesn't have to come from a surprise or unexpected result in the end. A lot of the stimulation of this subject comes from observing a pattern and asking 'Why?' That was the inspiration behind most of the chapters in this book.

Some interesting patterns like the ones discussed in the coincidence chapter can be the result of chance. Many others are there for a reason like buses coming in threes or petals coming in fives. Next time someone asks you what math is, don't say it is about learning your times tables. Math is the study of pretty patterns. And we all love pretty patterns.

Bibliography

The work of Martin Gardner, David Wells and Ian Stewart deserves special mention.

Gardner has written numerous books on recreational mathematics, of which *More Mathematical Puzzles & Diversions*, *Mathematical Circus*, and *Further Mathematical Diversions* were particularly useful references. All are published by Penguin.

David Wells has written two classic books on maths. A revised edition of *The Penguin Dictionary of Curious & Interesting Numbers* has recently been published, full of fascinating details behind every important number. *You are a Mathematician* is another David Wells book that was an invaluable source for us.

Ian Stewart the modern day Gardner, has contributed many popular maths articles to *Scientific American* and *New Scientist*. His book *Nature's Numbers* and Conway and Guy's *Book of Numbers* have more detailed discussions of the link between Fibonacci and plants.

We also recommend Darrell Huff's excellent book *How to Lie with Statistics* and David Kahn's *The codebreakers.* For more about Murphy's Law and why it comes true, read the delightful article by Robert J. Matthews, The Science of Murphy's Law (Scientific American, April 1997).

Other references that we used were:

The Complete Upmanship Stephen Potter

Silver Blaze Arthur Conan Doyle

Daisy, daisy, give me your answer do Ian Stewart, Scientific American, January 1995

Elementary cryptanalysis — a mathematical approach Abraham Sinkov, The Mathematical Association of America 1966

Penrose tiles to trapdoor ciphers Martin Gardner, Mathematical Association of America 1989

Management mathematics — a user friendly approach Peter Sprent, Penguin

A guide to operational research Eric Duckworth, Methuen & Co 1962

Time travel and other mathematical bewilderments Martin Gardner, W.H. Freeman & Company

Fair division: from cake cutting to dispute resolution Brams & Taylor, CUP

The classic cake problem Norman N Nelson and Forest N Fisch, The Mathematical Teacher Nov. 1973

Game theory —a non technical introduction Morton Davis, Basic Books

A compendium of math abuse from around the world AK Dewdney, Scientific American Nov. 1990

A partly true story Ian Stewart, Scientific American Feb 1993

Experiments in Topology Stephen Barr, John Murray
Fundamentals of operations research R.L. Ackoff & M.W. Sasieni, John Wiley & sons
Taxicab geometry offers a free ride to a Non-Euclid locale Martin Gardner, Scientific American Nov. 1980
Mathematics — exploring the world of numbers and space Irving Adler, Hamlyn
Encyclopaedia Britannica
For all Practical Purposes — Introducton to Contemporary Mathematics Various authors, WH Freeman
Mathematics — an introduction to its spirit and use (*readings from Scientific American*) Various authors, WH Freeman
Life Science Library — Mathematics David Berganini, Time Life Books

We would also like to thank the following who supplied us with data:

The International Tennis Federation (John Treleven), The Royal Institution, The Automobile Assoc, The RAC, Stephen Belcher, The Imperial War Museum, Ladbrokes, NOP, The Meteorological Office, The Office for National Statistics and Nielsen Media Research.

Index